ORDINARY DIFFERENTIAL EQUATIONS

African Institute of Mathematics Library Series

The African Institute of Mathematical Sciences (AIMS), founded in 2003 in Muizenberg, South Africa, provides a one-year postgraduate course in mathematical sciences for students throughout the continent of Africa. The **AIMS LIBRARY SERIES** is a series of short innovative texts, suitable for self-study, on the mathematical sciences and their applications in the broadest sense.

AIMS Library Series

ORDINARY DIFFERENTIAL EQUATIONS

A Practical Guide

BERND J. SCHROERS
Heriot-Watt University, Edinburgh

CAMBRIDGE
UNIVERSITY PRESS

CAMBRIDGE UNIVERSITY PRESS
Cambridge, New York, Melbourne, Madrid, Cape Town,
Singapore, São Paulo, Delhi, Tokyo, Mexico City

Cambridge University Press
The Edinburgh Building, Cambridge CB2 8RU, UK

Published in the United States of America by Cambridge University Press, New York

www.cambridge.org
Information on this title: www.cambridge.org/9781107697492

First published 2011

Printed in the United Kingdom at the University Press, Cambridge

A catalogue record for this publication is available from the British Library

ISBN 978-1-107-69749-2 Paperback

Contents

Preface

Purpose and scope

This book explains key concepts and methods in the field of ordinary differential equations. It assumes only minimal mathematical prerequisites but, at the same time, introduces the reader to the way ordinary differential equations are used in current mathematical research and in scientific modelling. It is designed as a practical guide for students and aspiring researchers in any mathematical science - in which I include, besides mathematics itself, physics, engineering, computer science, probability theory, statistics and the quantitative side of chemistry, biology, economics and finance.

The subject of differential equations is vast and this book only deals with initial value problems for ordinary differential equations. Such problems are fundamental in modern science since they arise when one tries to predict the future from knowledge about the present. Applications of differential equations in the physical and biological sciences occupy a prominent place both in the main text and in the exercises. Numerical methods for solving differential equations are not studied in any detail, but the use of mathematical software for solving differential equations and plotting functions is encouraged and sometimes required.

How to use this book

The book should be useful for students at a range of levels and with a variety of scientific backgrounds, provided they have studied differential and integral calculus (including partial derivatives), elements of real analysis (such as $\epsilon\delta$-definitions of continuity and differentiability), complex numbers and linear algebra. It could serve as a textbook for a first course on ordinary differential equations for undergraduates on a mathematics, science, engineering or economics degree who have studied the prerequisites listed above. For such readers it offers a fairly short and steep path through the material which includes vistas onto some advanced results. The book could also be used by more advanced undergraduates or early postgraduates who would like to revisit ordinary differential equations or who, studying them for the first time, want a compact treatment that takes into account their mathematical maturity.

Regardless of their background, readers should be aware that the chapters do not necessarily have to be studied in the order in which they are numbered. Having studied first order equations in Chapter 1, the reader can go on to study the theory underpinning general systems of differential equations in Chapter 2. Alternatively, if the mathematical generality of Chapter 2 seems too daunting, the reader could first study Chapter 3, which deals with specific second order equations and their applications in physics. This would mean taking certain recipes on trust, but would allow the reader to build up more experience with differential equations before mastering the general theory contained in Chapter 2.

The exercises form an integral part of the text, and the reader is strongly encouraged to attempt them *all*. They range in difficulty from routine to challenging, and some introduce or develop key ideas which are only mentioned in the main text. Solutions are not included, and it is quite likely that the reader will initially get stuck on some of the problems. This is entirely intended. Being stuck is an essential part of doing mathematics, and every mathematical training should include opportunities for getting stuck - and thus for learning how to get unstuck. One simple strategy is to move on and to return to the problem with a fresh mind later. Another is to discuss the problem with a fellow student - face to face or in 'cyberspace'.

The last chapter of the book consists of five projects. The first four are authentic in the sense that they address problems which arose in real and recent research. Readers, particularly those who have studied differential equations before, are encouraged to attempt (parts of) these projects before reaching the end of the book. In that way, the projects provide motivation for some of the theory covered in the book - and assure the reader that the techniques introduced are not mere finger exercises. If possible, the projects could be tackled by a small group of students working as a team. The last project is a guided proof of the existence and uniqueness theorem for initial value problems. The material there could also serve as the basis for a seminar on this proof.

Acknowledgements

This book grew out of courses on ordinary differential equations which I taught at Heriot-Watt University in Edinburgh and at the African Institute for Mathematical Sciences (AIMS) in Muizenberg, South Africa. In writing up my lecture notes I have tried to capture the spirit of the course at AIMS, with its strong emphasis on problem solving, team work and interactive teaching.

I am indebted to a number of people who contributed to this book. Ken Brown's notes on differential equations provided the foundations for my course on the subject at Heriot-Watt. Some of the pictures and type-set notes produced by Robert Weston for the same course appear, in recycled form, in Chapter 4. Aziz Ouhinou at AIMS led the seminar on the Picard-Lindelöf Theorem which evolved into Project 5. The students at AIMS whom I had the privilege of teaching during the past six years have profoundly shaped this book through their enthusiasm for learning, their mathematical intelligence and their probing questions. I am deeply grateful to them.

Notation and units

The mathematical notation in this book is mostly standard. The abbreviation 'iff' is occasionally used for 'if and only if'. The letter I denotes the identity matrix in Chapters 2 and 4. SI units are used for applications in physics, so distance is measured in metres (m), time in seconds (s), mass in kilograms (kg) and force in Newtons (N).

1
First order differential equations

1.1 General remarks about differential equations

1.1.1 Terminology

A differential equation is an equation involving a function and its derivatives. An example which we will study in detail in this book is the pendulum equation

$$\frac{d^2x}{dt^2} = -\sin(x), \tag{1.1}$$

which is a differential equation for a real-valued function x of one real variable t. The equation expresses the equality of two *functions*. To make this clear we could write (1.1) more explicitly as

$$\frac{d^2x}{dt^2}(t) = -\sin(x(t)) \qquad \text{for all } t \in \mathbb{R},$$

but this would lead to very messy expressions in more complicated equations. We therefore often suppress the independent variable.

When formulating a mathematical problem involving an equation, we need to specify *where* we are supposed to look for solutions. For example, when looking for a solution of the equation $x^2+1=0$ we might require that the unknown x is a real number, in which case there is no solution, or we might allow x to be complex, in which case we have the two solutions $x = \pm i$. In trying to solve the differential equation (1.1) we are looking for a twice-differentiable function $x : \mathbb{R} \to \mathbb{R}$. The set of all such functions is very big (bigger than the set of all real numbers, in a sense that can be made precise by using the notion of cardinality), and this is one basic reason why, generally speaking, finding solutions of differential equations is not easy.

Differential equations come in various forms, which can be classified as follows. If only derivatives of the unknown function with respect to one variable appear, we call the equation an *ordinary differential equation*, or ODE for short. If the function depends on several variables, and if partial derivatives with respect to at least two variables appear in the equation, we call it a *partial differential equation*, or PDE for short. In both cases, the order of the differential equation is the order of the highest derivative occurring in it. The independent variable(s) may be real or complex, and the unknown function may take values in the real or complex numbers, or in $\mathbb{R}^n$ of $\mathbb{C}^n$ for some positive integer n. In the latter case we can also think of the differential equation as a set of differential equations for each of the n components. Such a set is called a *system of differential equations of dimension* n. Finally we note that there may be parameters in a differential equation, which play a different role from the variables. The difference between parameters and variables is usually clear from the context, but you can tell that something is a parameter if no derivatives with respect to it appear in the equation. Nonetheless, solutions still depend on these parameters. Examples illustrating the terms we have just introduced are shown in Fig. 1.1.

In this book we are concerned with ordinary differential equations for functions of one real variable and, possibly, one or more parameters. We mostly consider real-valued functions but complex-valued functions also play an important role.

1.1.2 Approaches to problems involving differential equations

Many mathematical models in the natural and social sciences involve differential equations. Differential equations also play an important role in many branches of pure mathematics. The following system of ODEs for real-valued functions a, b and c of one

$$\frac{dy}{dx} - xy = 0$$

first order ordinary differential equation

$$\frac{du}{dt} = -v, \quad \frac{dv}{dt} = u$$

first order system of ordinary differential equations, dimension two

$$\frac{d^2y}{dx^2} = \left(\frac{dy}{dx}\right)^3$$

second order ordinary differential equation

$$\frac{1}{c^2}\frac{\partial^2 u}{\partial t^2} = \frac{\partial^2 u}{\partial x^2}$$

second order partial differential equation with parameter c

Fig. 1.1. Terminology for differential equations

real variable r arises in differential geometry:

$$\begin{aligned}
\frac{da}{dr} &= \frac{1}{2rc}(a^2 - (b-c)^2) \\
\frac{db}{dr} &= \frac{b}{2acr}(b^2 - (a-c)^2) \\
\frac{dc}{dr} &= \frac{1}{2ra}(c^2 - (a-b)^2),
\end{aligned}$$

One is looking for solutions on the interval $[\pi, \infty)$ subject to the initial conditions

$$a(\pi) = 0, \quad b(\pi) = \pi, \quad c(\pi) = -\pi.$$

At the end of this book you will be invited to study this problem as a part of an extended project. At this point, imagine that a colleague or friend had asked for your help in solving the above equations. What would you tell him or her? What are the questions that need to be addressed, and what methods do you know for coming up with answers? Try to write down some ideas before looking at the following list of issues and approaches.

(a) **Is the problem well-posed?** In mathematics, a problem is called well-posed if it has a solution, if that solution is unique and if the solution depends continuously on the data given in the problem, in a suitable sense. When a differential equation has one solution, it typically has infinitely many. In order to obtain a well-posed problem we therefore need to complement the differential equation with additional requirements on the solution. These could be initial conditions (imposed on the solution and its derivatives at one point) or boundary conditions (imposed at several points).

(b) **Are solutions stable?** We would often like to know how a given solution changes if we change the initial data by a small amount.

(c) **Are there explicit solutions?** Finding explicit solutions in terms of standard functions is only possible in rare lucky circumstances. However, when an explicit formula for a general solution can be found, it usually provides the most effective way of answering questions related to the differential equation. It is therefore useful to know the types of differential equation which can be solved exactly.

(d) **Can we find approximate solutions?** You may be able to solve a simpler version of the model exactly, or you may be able to give an approximate solution of the differential equation. In all approximation methods it is important to have some control over the accuracy of the approximation.

(e) **Can we use geometry to gain qualitative insights?** It is often possible to derive general, qualitative features of solutions without solving the differential equation. These could include asymptotic behaviour (what happens to the solution for large r?) and stability discussed under (b).

(f) **Can we obtain numerical solutions?** Many numerical routines for solving differential equations can be downloaded from open-source libraries like SciPy. Before using them, check if the problem you are trying to solve is well-posed. Having some insight into approximate or qualitative features of the solution usually helps with the numerical work.

In this text we will address all of these issues. We begin by looking at first order differential equations.

1.2 Exactly solvable first order ODEs

1.2.1 Terminology

The most general first order ODE for a real-valued function x of one real variable t is of the form

$$F\left(t, x, \frac{dx}{dt}\right) = 0, \tag{1.2}$$

for some real-valued function F of three variables. The function x is a solution if it is defined at least on some interval $I \subset \mathbb{R}$ and if

$$F\left(t, x(t), \frac{dx}{dt}(t)\right) = 0 \text{ for all } t \in I.$$

We call a first order ODE *explicit* if it can be written in terms of a real-valued function f of two variables as

$$\frac{dx}{dt} = f(t, x). \tag{1.3}$$

Otherwise, the ODE is called *implicit*. Before we try to understand the general case, we consider some examples where solutions can be found by elementary methods.

1.2.2 Solution by integration

The simplest kind of differential equation can be written in the form

$$\frac{dx}{dt} = f(t),$$

where f is a real-valued function of one variable, which we assume to be continuous. By the fundamental theorem of calculus we can find solutions by integration

$$x(t) = \int f(t)dt.$$

The right hand side is an indefinite integral, which contains an arbitrary constant. As we shall see, solutions of first order differential equations are typically only determined up to an arbitrary constant. Solutions to first order ODEs which contain an arbitrary constant are called *general solutions*.

Exercise 1.1 Revise your integration by finding general solutions of the following ODEs. Where are these solutions valid?

$$\text{(i)}\ \frac{dx}{dt} = \sin(4t-3), \qquad \text{(ii)}\ \frac{dx}{dt} = \frac{1}{t^2-1}.$$

1.2.3 Separable equations

Let us find the general solution of the following slightly more complicated equation:

$$\frac{dx}{dt} = 2tx^2. \tag{1.4}$$

It can be solved by separating the variables, i.e., by bringing all the t-dependence to one side and all the x-dependence to the other, and integrating:

$$\int \frac{1}{x^2}dx = \int 2tdt,$$

where we need to assume that $x \neq 0$. Integrating once yields

$$-\frac{1}{x} = t^2 + c,$$

with an arbitrary real constant c. Hence, we obtain a one-parameter family of solutions of (1.4):

$$x(t) = -\frac{1}{t^2+c}. \tag{1.5}$$

In finding this family of solutions we had to assume that $x \neq 0$. However, it is easy to check that the constant function $x(t) = 0$ for all t also solves (1.4). It turns out that the example is typical of the general situation: in separating variables we may lose constant solutions, but these can be recovered easily by inspection. A precise formulation of this statement is given in Exercise 1.8, where you are asked to prove it using an existence and uniqueness theorem which we discuss in Section 1.3.

The general solution of a first order ODE is really the family of all solutions of the ODE, usually parametrised by a real number. A first order ODE by itself is therefore, in general, not a well-posed problem in the sense of Section 1.1 because it does not have a unique solution. In the elementary examples of ODEs in Exercise

1.1 and also in the ODE (1.4) we can obtain a well-posed problem if we impose an initial condition on the solution. If we demand that a solution of (1.4) satisfies $x(0) = 1$ we obtain the unique answer $x(t) = \frac{1}{1-t^2}$. If, on the other hand, we demand that a solution satisfies $x(0) = 0$ then the constant function $x(t) = 0$ for all t is the only possibility. The combination of a first order ODE with a specification of the value of the solution at one point is called an *initial value problem.*

The method of solving a first order ODE by separating variables works, at least in principle, for any equation of the form

$$\frac{dx}{dt} = f(t)g(x),$$

where f and g are continuous functions. The solution $x(t)$ is determined implicitly by

$$\int \frac{1}{g(x)} dx = \int f(t) dt. \tag{1.6}$$

In general, it may be impossible to express the integrals in terms of elementary functions and to solve explicitly for x.

1.2.4 Linear first order differential equations

If the function f in equation (1.3) is a sum of terms which are either independent of x or linear in x, we call the equation linear. Consider the following example of an initial value problem for a linear first order ODE:

$$\frac{dx}{dt} + 2tx = t \qquad x(0) = 1. \tag{1.7}$$

First order linear equations can always be solved by using an *integrating factor.* In the above example, we multiply both sides of the differential equation by $\exp(t^2)$ to obtain

$$e^{t^2}\frac{dx}{dt} + 2te^{t^2}x = te^{t^2}. \tag{1.8}$$

Now the left hand side has become a derivative, and the equation (1.8) can be written as

$$\frac{d}{dt}\left(e^{t^2}x\right) = te^{t^2}.$$

Integrating once yields $xe^{t^2} = \frac{1}{2}e^{t^2} + c$, and hence the general solution

$$x(t) = \frac{1}{2} + ce^{-t^2}.$$

Imposing $x(0) = 1$ implies $c = \frac{1}{2}$ so that the solution of (1.7) is

$$x(t) = \frac{1}{2}\left(1 + e^{-t^2}\right).$$

General linear equations of the form

$$\frac{dx}{dt} + a(t)x = b(t), \tag{1.9}$$

where a and b are continuous functions of t, can be solved using the integrating factor

$$I(t) = e^{\int a(t)dt}. \tag{1.10}$$

Since the indefinite integral $\int a(t)dt$ is only determined up to an additive constant, the integrating factor is only determined up to a multiplicative constant: if $I(t)$ is an integrating factor, so is $C \cdot I(t)$ for any non-zero real number C. In practice, we make a convenient choice. In the example above we had $a(t) = 2t$ and we picked $I(t) = \exp(t^2)$. Multiplying the general linear equation (1.9) by $I(t)$ we obtain

$$\frac{d}{dt}\left(I(t)x(t)\right) = I(t)b(t).$$

Now we integrate and solve for $x(t)$ to find the general solution. In Section 2.5 we will revisit this method as a special case of the method of variation of the parameters.

1.2.5 Exact equations

Depending on the context, the independent variable in an ODE is often called t (particularly when it physically represents time), sometimes x (for example when it is a spatial coordinate) and sometimes another letter of the Roman or Greek alphabet. It is best not to get too attached to any particular convention. In the following example, the independent real variable is called x, and the real-valued function that we are looking for is called y. The differential equation governing y as a function of x is

$$(x + \cos y)\frac{dy}{dx} + y = 0. \tag{1.11}$$

Re-arranging this as

$$(x\frac{dy}{dx} + y) + \cos y\frac{dy}{dx} = 0,$$

we note that

$$(x\frac{dy}{dx} + y) = \frac{d}{dx}(xy), \quad \text{and} \quad \cos y\frac{dy}{dx} = \frac{d}{dx}\sin y.$$

If we define $\psi(x, y) = xy + \sin y$ then (1.11) can be written

$$\frac{d}{dx}\psi(x, y(x)) = 0 \tag{1.12}$$

and is thus solved by

$$\psi(x, y(x)) = c, \tag{1.13}$$

for some constant c.

Equations which can be written in the form (1.12) for some function ψ are called *exact*. It is possible to determine whether a general equation of the form

$$a(x, y)\frac{dy}{dx}(x) + b(x, y) = 0, \tag{1.14}$$

for differentiable functions $a, b : \mathbb{R}^2 \to \mathbb{R}$, is exact as follows. Suppose equation (1.14) were exact. Then we should be able to write it in the form (1.12) for a twice-differentiable function $\psi : \mathbb{R}^2 \to \mathbb{R}$. However, by the chain rule,

$$\frac{d}{dx}\psi(x, y(x)) = \frac{\partial\psi}{\partial y}\frac{dy}{dx}(x) + \frac{\partial\psi}{\partial x},$$

so that (1.14) is exact if there exists a twice-differentiable function $\psi : \mathbb{R}^2 \to \mathbb{R}$ with

$$\frac{\partial\psi}{\partial y}(x, y) = a(x, y) \quad \text{and} \quad \frac{\partial\psi}{\partial x}(x, y) = b(x, y).$$

Since $\dfrac{\partial^2\psi}{\partial x\partial y} = \dfrac{\partial^2\psi}{\partial y\partial x}$ for twice-differentiable functions we obtain a necessary condition for the existence of the function ψ:

$$\frac{\partial a}{\partial x} = \frac{\partial b}{\partial y}. \tag{1.15}$$

The equation (1.11) is exact because $a(x, y) = x + \cos y$ and $b(x, y) = y$ satisfy (1.15):

$$\frac{\partial}{\partial x}(x + \cos y) = 1 = \frac{\partial y}{\partial y}.$$

To find the function ψ systematically, we need to solve

$$\frac{\partial \psi}{\partial y} = x + \cos y, \qquad \frac{\partial \psi}{\partial x} = y. \tag{1.16}$$

As always in solving simultaneous equations, we start with the easier of the two equations; in this case this is the second equation in (1.16), which we integrate with respect to x to find $\psi(x, y) = xy + f(y)$, where f is an unknown function. To determine it, we use the first equation in (1.16) to derive $f'(y) = \cos y$, which is solved by $f(y) = \sin y$. Thus

$$\psi(x, y) = xy + \sin y,$$

leading to the general solution given in (1.13).

It is sometimes possible to make a non-exact equation exact by multiplying with a suitable integrating factor. However, it is only possible to give a recipe for computing the integrating factor in the linear case. In general one has to rely on inspired guesswork.

1.2.6 Changing variables

Some ODEs can be simplified and solved by changing variables. We illustrate how this works by considering two important classes. **Homogeneous ODEs** are equations of the form

$$\frac{dy}{dx} = f\left(\frac{y}{x}\right) \tag{1.17}$$

such as

$$\frac{dy}{dx} = \frac{2xy}{x^2 + y^2} = \frac{2(\frac{y}{x})}{1 + (\frac{y}{x})^2}. \tag{1.18}$$

If we define $u = \frac{y}{x}$, then $y = xu$ and thus

$$\frac{dy}{dx} = u + x\frac{du}{dx}.$$

Hence the equation (1.17) can be rewritten as

$$x\frac{du}{dx} = f(u) - u, \tag{1.19}$$

which is separable. In our example (1.18) we obtain the following equation for u:

$$x\frac{du}{dx} = \frac{2u}{1+u^2} - u = \frac{u(1-u^2)}{1+u^2}.$$

Provided $x \neq 0, u \neq 0$ and $u \neq \pm 1$ we can separate variables to obtain

$$\int \frac{1+u^2}{u(1-u)(1+u)} du = \int \frac{1}{x} dx.$$

Using partial fractions

$$\frac{1+u^2}{u(1-u)(1+u)} = \frac{1}{u} + \frac{1}{1-u} - \frac{1}{1+u}$$

we integrate to find

$$\ln \left| \frac{u}{1-u^2} \right| = \ln|x| + \tilde{c} \Leftrightarrow \frac{u}{1-u^2} = \pm cx, \tag{1.20}$$

where $c = \exp(\tilde{c})$ is a non-zero constant. By inspection we find that $u(x) = 0$ for all $x \in \mathbb{R}$ and $u(x) = \pm 1$ for all $x \in \mathbb{R}$ are also solutions. We can include the zero solution and absorb the sign ambiguity in (1.20) by allowing the constant c to take arbitrary real values. However, the solutions $u(x) = \pm 1$ have to be added to obtain the most general solution (compare again Exercise 1.8). The general solution of (1.19) is thus given by

$$y(x) = c(x^2 - y^2) \quad \text{or} \quad y(x) = x \quad \text{or} \quad y(x) = -x. \tag{1.21}$$

Note that these are well-defined for $x = 0$, and that they satisfy the original differential equation (1.18) there.

Bernoulli equations are of the form

$$\frac{dy}{dx} + a(x)y = b(x)y^\alpha,$$

where α is a real number not equal to 1. They are named after Jacob Bernoulli (1654–1705) who also invented the method for solving separable equations. The 'difficult' term here is the right hand side, so we divide by y^α (making a mental note that y is assumed to be non-vanishing) to find

$$y^{-\alpha}\frac{dy}{dx} + a(x)y^{1-\alpha} = b(x). \tag{1.22}$$

The equation simplifies when written in terms of

$$u = y^{1-\alpha}. \tag{1.23}$$

Using

$$\frac{du}{dx} = (1-\alpha)y^{-\alpha}\frac{dy}{dx},$$

the equation (1.22) becomes

$$\frac{1}{1-\alpha}\frac{du}{dx} + a(x)u = b(x).$$

This is a linear ODE and can be solved by finding an integrating factor. For example the equation

$$\frac{dy}{dx} + y = y^4$$

turns into a linear equation for $u = y^{-3}$

$$\frac{du}{dx} - 3u = -3,$$

which we can solve using the integrating factor $\exp(-3x)$. After integrating, do not forget to find the function y by inverting the relation (1.23).

Exercise 1.2 Name the type and find the general solution for each of the following first order equations:

(i) $\dfrac{dy}{dx} = \dfrac{e^x}{3+6e^x}$, (ii) $\dfrac{dy}{dx} = \dfrac{x^2}{y}$,

(iii) $\dfrac{dy}{dx} + 3y = x + e^{-2x}$, (iv) $x\dfrac{dy}{dx} = x\cos(2x) - y$,

(v) $\dfrac{dy}{dx} = \dfrac{y^2+2xy}{x^2}$, (vi) $xy^2 - x + (x^2y+y)\dfrac{dy}{dx} = 0$.

Exercise 1.3 Solve the following initial value problems.

$$\text{(i) } (\sin x + x^2 e^y - 1)\frac{dy}{dx} + y\cos x + 2xe^y = 0, \;\; y(0) = 0,$$

$$\text{(ii) } \frac{dx}{dt} + x = x^4, \;\; x(0) = 1, \;\; \text{(iii) } x\frac{dy}{dx} + y = x^4 y^3, \;\; y(1) = 1.$$

Exercise 1.4 Give equations for an ellipse, parabola and hyperbola in the (x, y)-plane and derive an exact differential equation for each.

Exercise 1.5 Consider the following mathematical model of epidemics. Assume that there is a community of N members with I infected and U uninfected individuals, so $U + I = N$. Define the ratios $x = I/N$ and $y = U/N$ and assume that N is constant and so large that x and y may be considered as continuous variables. Then we have $x, y \in [0, 1]$ and

$$x + y = 1. \tag{1.24}$$

Denoting time by t, the rate at which the disease spreads is $\frac{dx}{dt}$. If we make the assumption that the disease spreads by contact between sick and healthy members of the community, and if we further assume that both groups move freely among each other, we arrive at the differential equation

$$\frac{dx}{dt} = \beta xy, \tag{1.25}$$

where β is a real and positive constant of proportionality.

(i) Combine equations (1.24) and (1.25) to derive a differential equation for $x(t)$.

(ii) Find the solution of this differential equation for $x(0) = x_0$.

(iii) Show that $\lim_{t\to\infty} x(t) = 1$ if $x_0 > 0$ and interpret this result.

(iv) Is the model of epidemics studied here realistic? If not, what is missing in it?

Exercise 1.6 Two friends sit down to enjoy a cup of coffee. As soon as the coffee is poured, one of them adds cold milk to her coffee. The friends then chat without drinking any coffee. After five minutes the second friend also adds milk to her coffee, and both begin to drink. Which of them has got the hotter coffee? *Hint*: You may assume that the milk is colder than room temperature. If you do not know how to start this question look up Newton's law of cooling!

1.3 Existence and uniqueness of solutions

Most differential equations cannot be solved by the elementary methods described so far. There exist further techniques and tricks for finding explicit solutions of first order ODEs, but if you write down an example at random your chances of finding an explicit solution in terms of elementary functions are slim - even if you use all the tricks known to mathematicians. However, the mere absence of an explicit formula does not imply that no solution exists. Therefore, one would still like to have a general criterion to determine if a given ODE has a solution and also what additional conditions are needed to make sure that the solution is unique. Our experience so far suggests that solutions of first order ODEs are unique once we specify the value of the unknown function at one point. We could therefore ask when the general initial value problem

$$\frac{dx}{dt} = f(t,x), \qquad x(t_0) = x_0, \tag{1.26}$$

has a unique solution. An important theorem by Charles Emile Picard (1856–1941) and Ernst Lindelöf (1870–1946) says that, under fairly mild assumption on f, initial value problems have unique solutions, at least locally. Before we state the theorem we specify what we mean by a solution.

Definition 1.1 Let I be an interval and $t_0 \in I$. We say that a differentiable function $x : I \to \mathbb{R}$ is a solution of (1.26) in the interval I if $\frac{dx}{dt} = f(t,x)$ for all $t \in I$ and $x(t_0) = x_0$.

In this book we will use the following version of the Picard–Lindelöf Theorem:

Theorem 1.1 *Consider the intervals* $I_T = [t_0 - T, t_0 + T]$ *and* $B_d = [x_0 - d, x_0 + d]$ *for positive, real numbers* T, d. *Suppose that* $f : I_T \times B_d \to \mathbb{R}$ *is continuous and that its partial derivative*

$\frac{\partial f}{\partial x} : I \times B \to \mathbb{R}$ *is also continuous. Then there is a* $\delta > 0$ *so that the initial value problem* (1.26) *has a unique solution in the interval* $I_\delta = [t_0 - \delta, t_0 + \delta]$.

This version of the Picard–Lindelöf Theorem is useful in practice, but not the most general formulation. One can replace the requirement that the partial derivative $\frac{\partial f}{\partial x}$ exists and is continuous by the weaker condition that f satisfies a Lipschitz condition with respect to its second argument (see Definition 5.3 in the Project 5.5). The proof of the theorem proceeds via the re-formulation of the initial value problem (1.26) in terms of an integral equation, and the application of a contraction mapping theorem. We will not discuss the proof in the main part of this book, but in Project 5.5 you are guided through the main steps.

We would like to gain some understanding of how the Picard–Lindelöf Theorem works. To achieve this, we first sketch the various intervals and points mentioned in the theorem in the (t, x)-plane, leading to a picture like the one shown in Fig. 1.2. The diagonally shaded rectangle is the region $I_T \times B_d$ where the functions f and $\frac{\partial f}{\partial x}$ are continuous, i.e., where the conditions of the theorem are satisfied. The point (t_0, x_0) is contained in this region, and the graph of the solution we seek has to pass through this point (since $x(t_0) = x_0$). The smaller interval I_δ cuts a slice out of this rectangle, shown as the cross-hatched region. This slice is the region where we expect to have a unique solution of the initial value problem. The graph of such a solution is also sketched in the picture.

Now that we have a picture in mind, we are going to discuss the kind of questions that a mathematician might ask when confronted with the statement of an unfamiliar theorem. Is the meaning of each of the terms in the theorem clear? What happens when conditions of the theorem are violated? Can we give examples of initial value problems which do not satisfy the conditions of the Picard–Lindelöf Theorem and for which there are no or perhaps many solutions? These questions are addressed in the following remarks.

(i) Although the Picard–Lindelöf Theorem guarantees the existence of a solution for a large class of equations, it is not always

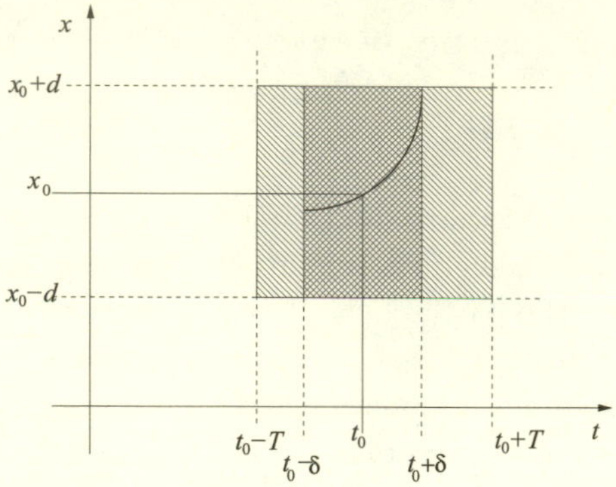

Fig. 1.2. A picture to illustrate the contents of the Picard–Lindelöf Theorem. Explanation in the main text

possible to find an explicit form of the solution in terms of elementary functions. For example

$$\frac{dx}{dt} = \sin(xt), \qquad x(0) = 1,$$

satisfies the condition of the theorem but no explicit formula is known for the solution.

(ii) If f is not continuous, then (1.26) may not have a solution. Consider, for example,

$$f(t,x) = \begin{cases} 1 & \text{if} \quad t \geq 0 \\ 0 & \text{if} \quad t < 0 \end{cases} \qquad \text{and} \quad x(0) = 0.$$

This would imply that $x(t) = t$ for $t \geq 0$ and $x(t) = 0$ for $t < 0$, which is not differentiable at $t = 0$. This is illustrated on the left in Fig. 1.3.

(iii) If f does not have a continuous first order partial derivative, (1.26) may have more than one solution. For example

$$\frac{dx}{dt} = x^{\frac{1}{3}}, \qquad x(0) = 0 \tag{1.27}$$

is solved by

$$x(t) = \begin{cases} (\frac{2}{3}(t-c))^{\frac{3}{2}} & \text{for} \quad t \geq c \\ 0 & \text{for} \quad t < c \end{cases}$$

for any $c \geq 0$. An example of such a solution is sketched in the middle in Fig. 1.3.
(iv) The Picard–Lindelöf Theorem guarantees a solution for t close to t_0, but this solution may not exist for all $t \in \mathbb{R}$. As we saw after (1.4), the initial value problem

$$\frac{dx}{dt} = 2tx^2, \qquad x(0) = 1$$

has the solution

$$x(t) = \frac{1}{1-t^2},$$

which tends to infinity as $t \to \pm 1$. Therefore, a solution in the sense of Definition 1.3 only exists in the interval $(-1, 1)$. This solution is shown on the right in Fig. 1.3. It is possible to determine the smallest interval in which a solution is guaranteed to exist - see the discussion in Section 5.5.
(v) Very straightforward-looking initial value problems may have no solution:

$$t\frac{dx}{dt} + x = t, \qquad x(0) = 1.$$

Setting $t = 0$ on both sides of the ODE we deduce that $x(0) = 0$, which is not consistent with the initial condition $x(0) = 1$. Why does the Picard–Lindelöf Theorem not apply?

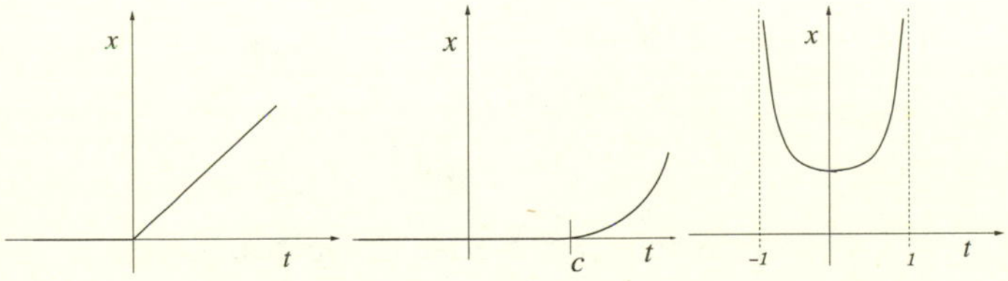

Fig. 1.3. Graphs illustrating the remarks (ii) (left), (iii) (middle) and (iv) (right) about the Picard–Lindelöf Theorem

Exercise 1.7 Show that each of the following initial value problems has two distinct solutions.
(i) $\frac{dx}{dt} = |x|^{1/2}$, $x(0) = 0$, (ii) $x\frac{dx}{dt} = t$, $x(0) = 0$.
Explain, in each case, why the hypotheses of the Picard–Lindelöf Theorem are not satisfied.

Exercise 1.8 Justify the method of separating variables as follows. Consider the first order differential equation

$$\frac{dx}{dt} = f(t)g(x), \tag{1.28}$$

where $f, g : \mathbb{R} \to \mathbb{R}$ are continuously differentiable functions. Show that if $g(\eta) = 0$, then the constant function $x(t) = \eta$ for all $t \in \mathbb{R}$ is a solution of (1.28). Use the Picard–Lindelöf Theorem to deduce that any solution x of (1.28) which satisfies $g(x(t_0)) \neq 0$ at some point t_0 must satisfy $g(x(t)) \neq 0$ for all t for which x is defined. Explain in which sense this result justifies the method of separation of variables.

Exercise 1.9 Curve sketching revision: sketch graphs of the following functions:
(i) $f(x) = \frac{1}{2}x^2 - 3x + \frac{9}{2}$, (ii) $f(x) = \sqrt{x+2}$,
(iii) $f(x) = 3\cos(\pi x + 1)$, (iv) $f(x) = -e^{(x-1)}$,
(v) $f(x) = \ln(x+4)$ (vi) The solutions (1.21), for a range of values for c.

Exercise 1.10 Find solutions of each of the equations

$$\text{(i)}\ \frac{dx}{dt} = x^{\frac{1}{2}}, \quad \text{(ii)}\ \frac{dx}{dt} = x, \quad \text{(iii)}\ \frac{dx}{dt} = x^2,$$

satisfying the initial condition $x(0) = 1$, and sketch them. Comment on the way solutions of $\dfrac{dx}{dt} = x^\alpha$ grow in each of the cases (i) $0 < \alpha < 1$, (ii) $\alpha = 1$, (iii) $\alpha > 1$.

1.4 Geometric methods: direction fields

In this section we explain how to sketch the graph of solutions of first order ODEs *without* solving the ODE. The tool that allows one to do this is called the *direction field*. To understand how it works we first assume that we have a family of solutions and compute the associated direction field. Then we reverse the process and sketch the solutions directly from the direction field.

Consider the general form of an explicit first order ODE

$$\frac{dx}{dt}(t) = f(t, x(t)), \tag{1.29}$$

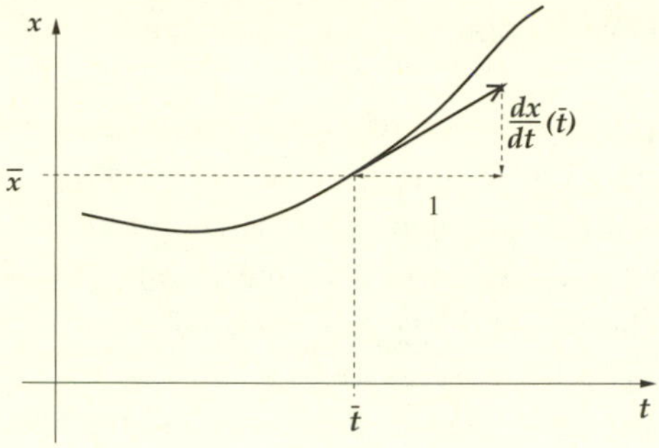

Fig. 1.4. Tangent to the graph of a function $x(t)$

where we have exhibited the independent variable t for the sake of the following discussion. Suppose we have a solution $x(t)$ of (1.29) and draw its graph in the (t, x)-plane, as shown in Fig. 1.4. Suppose that $x(\bar{t}) = \bar{x}$, i.e., the point $(\bar{t}, \bar{x})$ lies on the graph of $x(t)$. Then the slope of the graph at $(\bar{t}, \bar{x})$ is $\dfrac{dx}{dt}(\bar{t})$. Geometrically, this means that if we start at $(\bar{t}, \bar{x})$ and move one step in the t-direction we have to move $\dfrac{dx}{dt}(\bar{t})$ steps in the x-direction in order to stay on the tangent to the curve through $(\bar{t}, \bar{x})$. In other words, the direction of the tangent at (t, x) is given by the vector

$$\begin{pmatrix} 1 \\ \dfrac{dx}{dt}(\bar{t}) \end{pmatrix}.$$

Now note that (1.29) allows us to replace $\dfrac{dx}{dt}(\bar{t})$ by $f(\bar{t}, \bar{x})$. Thus we can write down the tangent vector to the solution curve - without knowing the solution! This motivates the definition of the *direction field* of the differential equations (1.29) as the map

$$\boldsymbol{V} : \mathbb{R}^2 \to \mathbb{R}^2, \qquad \boldsymbol{V}(t, x) = \begin{pmatrix} 1 \\ f(t, x) \end{pmatrix}. \tag{1.30}$$

It is a collection of all tangent vectors to graphs of all possible solutions of the original ODE (1.29). The direction field is an example of a *vector field*: a map which assigns a vector (in our case in $\mathbb{R}^2$) to every point in a set (in our case this happens to be $\mathbb{R}^2$, too).

The graph of any solution of (1.29) through a point (t, x) must have the tangent vector $\boldsymbol{V}(t, x)$ given in (1.30). To understand the general nature of all solutions it is therefore helpful to sketch the vector field $\boldsymbol{V}$, i.e., to pick a few points $(t_1, x_1), (t_2, x_2), \ldots$ and to draw the vector $\boldsymbol{V}(t_1, x_1)$ with base at (t_1, x_1), then the vector $\boldsymbol{V}(t_2, x_2)$ with base at (t_2, x_2) and so on. It is useful to think of the direction field as 'signpostings' for someone walking in the (t, x)-plane: by following the arrows at every point in the plane the walker's path ends up being the graph of a solution of the differential equation (1.29). This point of view also provides a very important way of thinking about a solution of an ODE: it is a global path which is determined by lots of local instructions.

Example 1.1 Sketch the direction field for the ODE

$$\frac{dx}{dt} = t^2 + x^2. \tag{1.31}$$

Hence sketch the solutions satisfying the initial conditions
(i) $x(-1) = -1$, (ii) $x(-1) = 0$.

In order to sketch the direction field we evaluate it at a couple of points, e.g.,

$$\boldsymbol{V}(0,0) = \begin{pmatrix} 1 \\ 0 \end{pmatrix}, \quad \boldsymbol{V}(1,0) = \begin{pmatrix} 1 \\ 1 \end{pmatrix}, \quad \boldsymbol{V}(0,1) = \begin{pmatrix} 1 \\ 1 \end{pmatrix},$$
$$\boldsymbol{V}(1,1) = \begin{pmatrix} 1 \\ 2 \end{pmatrix}, \boldsymbol{V}(-1,1) = \begin{pmatrix} 1 \\ 2 \end{pmatrix}, \quad \boldsymbol{V}(2,0) = \begin{pmatrix} 1 \\ 4 \end{pmatrix}.$$

We draw a horizontal arrow representing $\begin{pmatrix} 1 \\ 0 \end{pmatrix}$ starting at the origin, arrows representing $\begin{pmatrix} 1 \\ 1 \end{pmatrix}$ starting at the points $(1, 0)$ and $(0, 1)$ in the (t, x)-plane, the arrow for $\begin{pmatrix} 1 \\ 2 \end{pmatrix}$ at $(1, 1)$ and so on, leading to the picture shown in Fig. 1.5.

From the formula for the direction field it is clear that vectors attached to all points on a circle $t^2 + x^2 =$ const. are the same, and that the arrows point up more steeply the further away we are

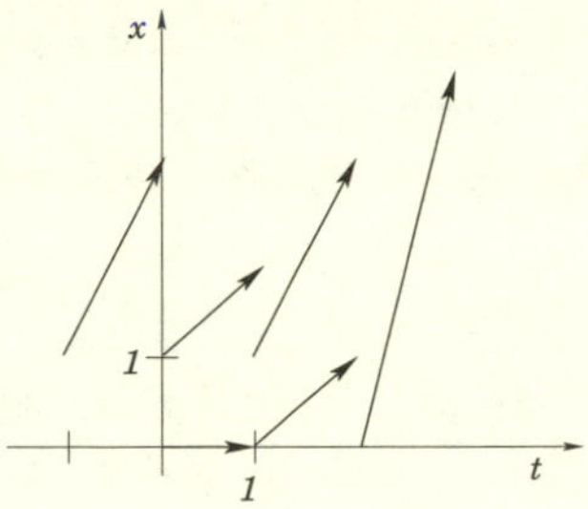

Fig. 1.5. Sketching the direction field for the differential equation (1.31)

from the origin. Generally, curves on which the direction field is constant are called *isoclines* and are helpful in sketching direction fields. Recalling that our interest in the direction field stems from the fact that it is tangent to any solution we notice that we can rescale the length of the vectors by a constant without affecting this property. It is therefore often better to make all the vectors shorter by a suitably chosen constant in order to avoid a crowded picture. This was done in producing Fig. 1.6, which also shows the required solution curves. □

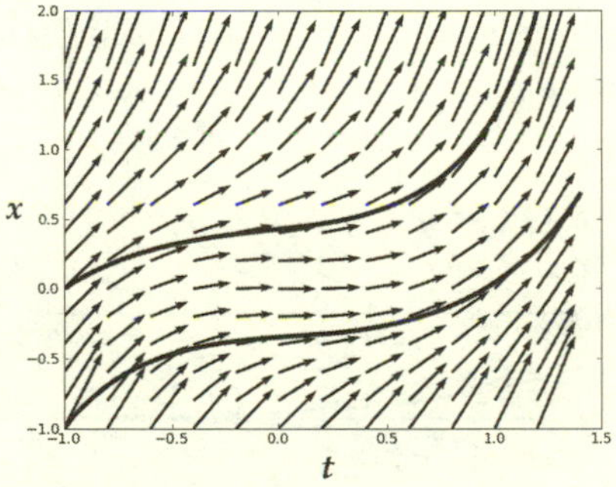

Fig. 1.6. Direction field and graphs of solutions for the differential equation (1.31)

It is not always possible to figure out the behaviour of all solutions from the direction field. However, the following rule is often very helpful.

Lemma 1.1 *If the function $f : \mathbb{R}^2 \to \mathbb{R}$ satisfies the conditions of Picard–Lindelöf Theorem in a region $U \in \mathbb{R}^2$, it is impossible for graphs of solutions of the differential equation* (1.29) *to cross in the region U.*

Proof If graphs of two solutions $x, \tilde{x}$ were to cross in a point (t_0, x_0), then the initial value problem (1.29) with $x(t_0) = x_0$ would have two solutions, which contradicts the uniqueness of the solution guaranteed by the Picard–Lindelöf Theorem. □

Example 1.2 Sketch the direction fields for the differential equation

$$\frac{dx}{dt} = x(x-1)(x-2). \tag{1.32}$$

Find all constant solutions and hence sketch the solutions with the initial conditions (i) $x(0) = 0.5$, (ii) $x(0) = 1.5$.

In this case the vectors representing (1.30) do not depend on t. They point down for $x < 0$ and $1 < x < 2$, up for $0 < x < 1$ and $x > 2$, and are horizontal for $x = 0, x = 1, x = 2$. In sketching, we again rescale by a suitable factor. Next we note that the equation has the constant solutions $x(t) = 0$ for all t, $x(t) = 1$ for all t, $x(t) = 2$ for all t. We thus draw horizontal lines at $x = 0$, $x = 1$ and $x = 2$. The graphs of other solutions must not cross these lines. Hence we obtain sketches like those shown in Fig. 1.7. □

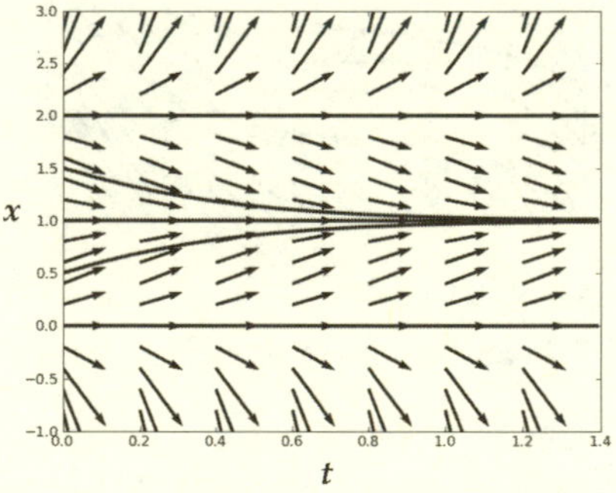

Fig. 1.7. Direction field and graphs of solutions for the differential equation (1.32)

Exercise 1.11 Find all constant solutions of $\frac{dx}{dt} + x^2 = 1$ and sketch the direction field for this equation. Hence sketch the solutions satisfying
(i) $x(0) = 1.1$, (ii) $x(0) = 0$, (iii) $x(0) = -1.1$.

Exercise 1.12 Consider the ODE $\frac{dx}{dt} = x^2 - t$. Determine the points in the (t, x)-plane at which

$$\text{(i)}\,\frac{dx}{dt} = 0,\ \text{(ii)}\,\frac{dx}{dt} > 0,\ \text{and (iii)}\,\frac{dx}{dt} < 0.$$

Hence sketch the direction field. Also sketch the solution satisfying $x(1) = 1$.

1.5 Remarks on numerical methods

At various points in this book you will be asked to use mathematical software for solving ODEs on a computer. When first studying ODEs it is not unreasonable to treat computer programmes for solving ODEs largely as a 'black box': you learn which input the programme requires, provide the input data and then take the programme's output on trust. This is the way most people use pocket calculators when evaluating $\sqrt{7}$, for example. However, as mathematicians we would like to have some general understanding of what is inside the 'black box' - just like we would have some idea of how to compute approximations to $\sqrt{7}$.

In this short section we develop a general notion of what is involved in solving an initial value problem numerically. If you use numerical ODE solvers extensively in your work you will need to consult other sources to learn more about the different algorithms for solving ODEs numerically. In particular, you will need to find out which algorithms are well adapted to solving the kind of ODEs you are interested in. However, it is worth stressing that numerical methods, even when carefully selected, do not solve all problems one encounters in studying ODEs. For example, computers are of limited use in solving ODEs with singularities or for understanding qualitative properties of a large class of solutions. Both of these points are illustrated in Project 5.4 at the end of this book.

Most methods for solving ODEs numerically are based on the

geometrical way of thinking about initial value problems discussed after (1.29). We saw there that the initial value problem

$$\frac{dx}{dt} = f(t,x), \qquad x(t_0) = x_0,$$

can be thought of as a set of instructions for moving in the (t,x)-plane, starting from the point (t_0, x_0) specified by the initial condition. Numerical algorithms for initial value problems essentially construct the solution by following these instructions step by step. However, knowing a solution $x(t)$ even on a small interval I would require recording the value of $x(t)$ at infinitely many points $t \in I$. No computer can store infinitely many data points, so numerical algorithms for solving ODEs invariably require a discretisation of the problem: a finite number of points $t_i, i = 0, \ldots n$, in I is selected, and the algorithm will return values x_i which are good approximations to the values $x(t_i)$ of the true solution evaluated at the points t_i.

Suppose we want to know the values of a solution $x(t)$ of (1.29) at equally spaced points $t_0, t_1, \ldots, t_n$,with $t_{i+1} - t_i = h$, given that $x(t_0) = x_0$. Since the function f in the ODE (1.29) is given, we can compute $\frac{dx}{dt}(t_0) = f(t_0, x_0)$. Thus we have the first two terms in the Taylor expansion of the solution

$$x(t_1) = x(t_0 + h) = x(t_0) + hf(t_0, x_0) + \ldots,$$

where we neglected terms which are of quadratic or higher order in h. If h is sufficiently small, the number

$$x_1 = x_0 + hf(t_0, x_0)$$

is therefore a good approximation to $x(t_1)$. Then we compute an approximation x_2 to $x(t_2)$ by repeating the procedure, i.e.,

$$x_2 = x_1 + hf(t_1, x_1).$$

The iteration of this step is the *Euler method* for computing numerical approximations x_i to the values $x(t_i)$ of a solution of (1.29). The general formula of the Euler method is

$$x_{i+1} = x_i + hf(t_i, x_i), \quad i = 0, \ldots, n-1. \tag{1.33}$$

To illustrate this scheme, consider the simple initial value problem

$$\frac{dx}{dt} = x, \quad x(0) = 1.$$

We can solve this exactly (by many of the methods in Section 1.2, or by inspection) and have the exact solution $x(t) = e^t$. Let us apply the Euler method, with $t_0 = 0, t_n = 1$ and $h = 1/n$. Then (1.33) gives the recursion

$$x_{i+1} = \left(1 + \frac{1}{n}\right) x_i,$$

which we easily solve to obtain

$$x_n = \left(1 + \frac{1}{n}\right)^n$$

as an approximation to the value of the exact solution at $t = 1$, i.e., to $x(1) = e$. Using the limit

$$\lim_{n\to\infty} \left(1 + \frac{1}{n}\right)^n = e,$$

we see that, in this case, the Euler method approximates the true solution evaluated at $t = 1$ arbitrarily well as $n \to \infty$ or, equivalently, as the stepsize $h = 1/n$ goes to zero.

In general, the Euler method is not the most effective way to obtain a good numerical approximation to solutions of ODEs. One way of improving the Euler method is based on the observation that the general form (1.26) of an initial value problem is equivalent to the equation

$$x(t) = x_0 + \int_{t_0}^{t} f(s, x(s))\, ds. \tag{1.34}$$

This kind of equation is called an *integral equation* for x since it contains the unknown function x and an integral involving x. You are asked to prove the equivalence with (1.26) in Exercise 1.13. Suppose now that the function f satisfies the condition of the Picard–Lindelöf Theorem on $\mathbb{R} \times \mathbb{R}$, so that (1.26) and therefore also (1.34) has a unique solution. We would like to compute good approximations x_i to the values of that solution at points $t_i = t_0 + ih$, for some (small) increment h and $i = 1, \ldots, n$. From (1.34) we can deduce the (exact) recursion relation

$$x(t_{i+1}) = x(t_i) + \int_{t_i}^{t_{i+1}} f(s, x(s))\, ds. \tag{1.35}$$

The Euler recursion (1.33) is equivalent to approximating the integral in this formula via

$$\int_{t_i}^{t_{i+1}} f(s, x(s))\, ds \approx h f(t_i, x(t_i)),$$

which amounts to estimating the integral over each interval $[t_i, t_{i+1}]$ by the area of a rectangle of width h and height $f(t_i, x(t_i))$. A better approximation to the integral in (1.35) can be obtained by estimating the area in terms of the trapezium rule, which takes into account the value of the function $f(s, x(s))$ at both endpoints t_i and t_{i+1} of the interval. This would lead to the recursion

$$x_{i+1} = x_i + \frac{h}{2}\left(f(t_i, x_i) + f(t_{i+1}, x_{i+1})\right).$$

The trouble with this formula is that x_{i+1} appears on both sides. To get around this we approximate x_{i+1} on the right hand side by the (old) Euler method (1.33), thus obtaining the recursion

$$x_{i+1} = x_i + \frac{h}{2}\left(f(t_i, x_i) + f(t_{i+1}, x_i + h f(t_i, x_i))\right). \tag{1.36}$$

This scheme is called the improved Euler method or Heun's method.

Exercise 1.13 Prove the equivalence between the initial value problem (1.26) and the integral equation (1.34) on an interval $[t_0 - T, t_0 + T]$, for any $T > 0$.

Exercise 1.14 Solve the initial value problem $\frac{dx}{dt} = 0.5 - t + 2x$, $x(0) = 1$ numerically and compute approximations to the values of x at $t = 0.1$, $t = 0.5$ and $t = 1$ in three ways: (a) using the Euler method with $h = 0.05$, (b) using the improved Euler method with $h = 0.1$, (c) using the improved Euler method with $h = 0.05$. Also find the exact solution and compare the approximate with the exact values!

2
Systems and higher order equations

2.1 General remarks

Very few real-life problems can be modelled successfully in terms of a single, real-valued function of one real variable. Typically, we need to keep track of several quantities at the same time, and thus need several real-valued functions to construct a realistic mathematical model. In studying a chemical reaction, for example, we may want to know the concentrations of several reacting substances as functions of time. Also, to describe the motion of a single particle in space, we need three functions of time for the three position coordinates. If several particles are involved, we need three such functions per particle. In such cases, modelling the time evolution of all relevant quantities typically requires a system of ODEs.

To fix our notation, we denote the independent variable by t and the dependent functions by $x_1, \ldots, x_n$. Each of these functions obeys a differential equation, which may in turn depend on the other functions, leading to a system of differential equations. It could look like this:

$$\frac{dx_1}{dt} = tx_1^2 + \sin(x_2), \qquad \frac{dx_2}{dt} = x_1 - 4t^2x_2. \tag{2.1}$$

If we combine the functions $x_1, x_2 : \mathbb{R} \to \mathbb{R}$ into one vector-valued function

$$\boldsymbol{x} : \mathbb{R} \to \mathbb{R}^2, \quad \boldsymbol{x}(t) = \begin{pmatrix} x_1(t) \\ x_2(t) \end{pmatrix},$$

we can write the system of differential equations (2.1) as one equation for $\boldsymbol{x}$:

$$\frac{d\boldsymbol{x}}{dt} = \begin{pmatrix} tx_1^2 + \sin(x_2) \\ x_1 - 4t^2 x_2 \end{pmatrix}.$$

More generally, a system of ODEs of dimension n can be written in terms of the vector-valued function

$$\boldsymbol{x} : \mathbb{R} \to \mathbb{R}^n, \quad \boldsymbol{x}(t) = \begin{pmatrix} x_1(t) \\ \vdots \\ x_n(t) \end{pmatrix} \tag{2.2}$$

in the form

$$\frac{d\boldsymbol{x}}{dt} = \boldsymbol{f}(t, \boldsymbol{x}), \tag{2.3}$$

for a suitable function $\boldsymbol{f} : \mathbb{R}^{n+1} \to \mathbb{R}^n$. Although the general form (2.3) seems to involve only the first derivative, it includes differential equations of higher order as a special case. To show this, we use the abbreviations

$$\dot{x} = \frac{dx}{dt}, \quad \ddot{x} = \frac{d^2x}{dt^2}, \quad x^{(n)} = \frac{d^nx}{dt^n}. \tag{2.4}$$

Then an nth order differential equation of the form

$$x^{(n)}(t) = f(t, x, , \ldots, x^{(n-1)}) \tag{2.5}$$

can be written as a first order equation for the vector-valued function $\boldsymbol{x}$ in (2.2) as

$$\frac{d}{dt}\begin{pmatrix} x_1 \\ x_2 \\ \vdots \\ x_n \end{pmatrix} = \begin{pmatrix} x_2 \\ x_3 \\ \vdots \\ f(t, x_1, \ldots, x_n) \end{pmatrix}. \tag{2.6}$$

We recover the differential equation (2.5) by identifying $x = x_1$. Then the first $n - 1$ components of (2.6) tell us that

$$x_2 = \dot{x}, \ x_3 = \dot{x}_2 = \ddot{x}, \ \ldots, x_n = x^{(n-1)}$$

so that the last component is precisely the equation (2.5).

Systems of second order differential equations play an important role in physics because Newton's second law (Sir Isaac Newton,

1643–1727) for the motion of a body in space is a second order differential equation of the form

$$m\frac{d^2\boldsymbol{x}}{dt^2} = \boldsymbol{F}(t, \boldsymbol{x}), \tag{2.7}$$

where t is time, $\boldsymbol{x} : \mathbb{R} \to \mathbb{R}^3$ is the position vector of the body as a function of time and $\boldsymbol{F}(t, \boldsymbol{x})$ is the force acting on the body at time t and when the body is at position $\boldsymbol{x}$. The equation (2.7) is also often called the *equation of motion* for the body. We will study the equation of motion for a particular physical system in detail in Section 3.2.

Example 2.1 Write the following differential equations as first order systems.

$$\begin{aligned} &\text{(i)}\, \frac{d^2x}{dt^2} + \frac{dx}{dt} + 4x = 0, \quad \text{(ii)}\, x\ddot{x} + (\dot{x})^2 = 0, \\ &\text{(iii)}\, \ddot{\boldsymbol{x}} = -\frac{GM}{|\boldsymbol{x}|^3}\boldsymbol{x}, \quad G, M \text{ real constants.} \end{aligned} \tag{2.8}$$

For (i) we introduce a two-component vector $\boldsymbol{x} = \begin{pmatrix} x_1 \\ x_2 \end{pmatrix}$. We identify x_1 with x and want our equation to ensure that $x_2 = \dot{x} = \dot{x}_1$. The required equation is

$$\frac{d}{dt}\begin{pmatrix} x_1 \\ x_2 \end{pmatrix} = \begin{pmatrix} x_2 \\ -4x_1 - x_2 \end{pmatrix} = \begin{pmatrix} 0 & 1 \\ -4 & -1 \end{pmatrix}\begin{pmatrix} x_1 \\ x_2 \end{pmatrix}.$$

For (ii) we again introduce real functions x_1, x_2 and set $x = x_1$. The differential equation can then be written as

$$\dot{x}_1 = x_2, \quad x_1\dot{x}_2 + x_2^2 = 0.$$

Finally, the equation (iii) is the equation of motion (2.7) for a particle moving in the gravitational field of another body (e.g., a star) of mass M. It is a second order system for a vector-valued function $\boldsymbol{x} : \mathbb{R} \to \mathbb{R}^3$. If we introduce a second vector-valued function $\boldsymbol{v} : \mathbb{R} \to \mathbb{R}^3$, we can write it as a first order system for two vectors:

$$\dot{\boldsymbol{x}} = \boldsymbol{v}, \quad \dot{\boldsymbol{v}} = -\frac{GM}{|\boldsymbol{x}|^3}\boldsymbol{x}.$$

Since the vector-valued functions $\boldsymbol{x}$ and $\boldsymbol{v}$ have three components each, we obtain a first order system of dimension six. □

The possibility of writing *any* ODE as a *first order* system is important because it provides a universal framework for proving theorems and for numerical work. Once we have proved a theorem for a first order system we can immediately deduce corollaries for ODEs of higher order. Similarly, any numerical algorithm which solves first order systems can deal with higher order ODEs - we only need to translate the higher order ODE into a first order system. For this reason, it is very useful to have a standard numerical differential equation solver for systems easily available. Such solvers can be obtained, for example, as a part of the free, open-source software Sage or as functions in the free open-source library SciPy. Regardless of whether you use free or commercial software, you are strongly encouraged to familiarise yourself with one ODE solver now. Exercise 2.1 suggests a second order ODE that you might use as an example. Note, finally, that writing a higher order ODE as a first order system is usually *not* the most efficient way of finding exact solutions. This fact is illustrated in Exercise 2.2 and in many examples in Chapter 3.

Exercise 2.1 Write the pendulum equation (1.1) as a first order system. Hence use an ODE solver of any mathematical software to which you have access to find solutions of the pendulum equation satisfying the following initial conditions:
(i) $x(0) = 1.0, \dot{x}(0) = 0$, (ii) $x(0) = \pi, \dot{x}(0) = 1.0$.
Plot $x(t)$ for both solutions in the interval $t \in [0, 10)$.

Exercise 2.2 (Free fall) The force acting on a freely falling body of mass m near the Earth's surface has magnitude mg, where $g = 9.8$ m s^{-1}, and points downwards. In suitable coordinates, with the x- and y-axes horizontal and the z-axis pointing up, the equation of motion (2.7) is thus

$$m\frac{d^2}{dt^2}\begin{pmatrix} x \\ y \\ z \end{pmatrix} = \begin{pmatrix} 0 \\ 0 \\ -gm \end{pmatrix}. \tag{2.9}$$

Find the general solution of (2.9). It should contain six constants of integration. What is their physical significance?

2.2 Existence and uniqueness of solutions for systems

The Picard–Lindelöf Theorem 1.1 can be generalised to systems of ODEs. In our formulation, we use the notation

$$||\boldsymbol{x}|| = \sqrt{x_1^2 + \ldots x_n^2}$$

for the Euclidean length of vectors in $\mathbb{R}^n$ (see also Project 5.5).

Theorem 2.1 *Consider the closed interval $I_T = [t_0 - T, t_0 + T]$ and the closed ball $B_d = \{\boldsymbol{y} \in \mathbb{R}^n | \, ||\boldsymbol{y} - \boldsymbol{x}_0|| \leq d\}$ in $\mathbb{R}^n$, where T, d are positive, real numbers. Suppose that the function*

$$\boldsymbol{f} : I_T \times B_d \to \mathbb{R}^n$$

is continuous and that the partial derivatives in the matrix

$$D\boldsymbol{f} = \begin{pmatrix} \frac{\partial f_1}{\partial x_1} & \cdots & \frac{\partial f_1}{\partial x_n} \\ \vdots & & \vdots \\ \frac{\partial f_n}{\partial x_1} & \cdots & \frac{\partial f_n}{\partial x_n} \end{pmatrix} \tag{2.10}$$

exist and are continuous in $I_T \times B_d$. Then there exists a $\delta > 0$ so that the initial value problem

$$\frac{d\boldsymbol{x}}{dt}(t) = \boldsymbol{f}(t, \boldsymbol{x}), \quad \boldsymbol{x}(t_0) = \boldsymbol{x}_0 \tag{2.11}$$

has a unique solution in the interval $I_\delta = [t_0 - \delta, t_0 + \delta]$.

For the proof of this version of the Picard–Lindelöf Theorem we again refer the reader to Project 5.5. As with the simpler Theorem 1.1 for first order ODEs we would like to understand the theorem better. We begin by looking at its assumptions.

Example 2.2 Compute the derivative matrix (2.10) for

$$\text{(i) } \boldsymbol{f}(t, \boldsymbol{x}) = \begin{pmatrix} t & 4 \\ -4 & t^2 \end{pmatrix} \begin{pmatrix} x_1 \\ x_2 \end{pmatrix} + \begin{pmatrix} \cos(t) \\ 0 \end{pmatrix},$$

$$\text{(ii) } \boldsymbol{f}(\boldsymbol{x}) = \begin{pmatrix} x_2 - x_1 r \\ -x_1 - x_2 r \end{pmatrix}, \quad r = \sqrt{x_1^2 + x_2^2}. \tag{2.12}$$

Determine the maximal region where the conditions of the Picard–Lindelöf Theorem are satisfied.

The derivative matrix for example (i) is

$$Df(t, \boldsymbol{x}) = \begin{pmatrix} t & 4 \\ -4 & t^2 \end{pmatrix},$$

which is a continuous function of $t \in \mathbb{R}$ and independent of $\boldsymbol{x}$. Hence the conditions of the Picard–Lindelöf Theorem are satisfied on $\mathbb{R} \times \mathbb{R}^2$. The derivative matrix for example (ii) is

$$D\boldsymbol{f}(\boldsymbol{x}) = \begin{pmatrix} -r - \frac{x_1^2}{r} & 1 - \frac{x_1 x_2}{r} \\ -1 - \frac{x_1 x_2}{r} & -r - \frac{x_2^2}{r} \end{pmatrix}, \qquad (2.13)$$

where we used $\partial r/\partial x_1 = x_1/r$ etc. The functions in this matrix are continuous for all $\boldsymbol{x} \neq (0,0)$, but some appear to be ill-defined at the origin $(0,0)$. To work out the partial derivatives at the origin we have to go back to first principles. Noting

$$\frac{\partial f_1}{\partial x_1}(0,0) = \lim_{h \to 0} \frac{f_1(h,0) - f_1(0,0)}{h} = \lim_{h \to 0}(-\sqrt{|h|}) = 0,$$
$$\frac{\partial f_1}{\partial x_2}(0,0) = \lim_{h \to 0} \frac{f_1(0,h) - f_1(0,0)}{h} = \lim_{h \to 0}(1 - \sqrt{|h|}) = 1,$$
$$\frac{\partial f_2}{\partial x_1}(0,0) = \lim_{h \to 0} \frac{f_2(h,0) - f_2(0,0)}{h} = \lim_{h \to 0}(-1 - \sqrt{|h|}) = -1,$$
$$\frac{\partial f_2}{\partial x_2}(0,0) = \lim_{h \to 0} \frac{f_2(0,h) - f_2(0,0)}{h} = \lim_{h \to 0}(-\sqrt{|h|}) = 0,$$

we conclude

$$D\boldsymbol{f}(0,0) = \begin{pmatrix} 0 & 1 \\ -1 & 0 \end{pmatrix}. \qquad (2.14)$$

To see that this makes the partial derivatives continuous for all $\boldsymbol{x} \in \mathbb{R}^2$ we return to the functions appearing in (2.13). Using $|x_1 x_2| \leq \frac{1}{2}(x_1^2 + x_2^2)$ and $x_1^2 \leq (x_1^2 + x_2^2)$ as well as $x_2^2 \leq (x_1^2 + x_2^2)$ we compute the limits

$$\lim_{\boldsymbol{x} \to 0} \frac{|x_1 x_2|}{r} \leq \frac{1}{2} \lim_{\boldsymbol{x} \to 0} r = 0,$$

and

$$\lim_{\boldsymbol{x} \to 0} \frac{x_i^2}{r} \leq \lim_{\boldsymbol{x} \to 0} r = 0, \quad \text{for} \quad i = 1, 2.$$

Thus we have shown

$$\lim_{\boldsymbol{x} \to 0} D\boldsymbol{f}(\boldsymbol{x}) = D\boldsymbol{f}(0,0)$$

and that the conditions of the Picard–Lindelöf Theorem hold in $\mathbb{R} \times \mathbb{R}^2$. We have treated this example in some detail since we will revisit it in Exercises 2.5 and 4.27. □

The Picard–Lindelöf Theorem has immediate and important implications for n-th order ODEs. Consider the equation (2.5) for a single function x. As we saw, this can be formulated as a system of ODEs for an n-component vector $\boldsymbol{x}$ (2.6). The right hand side of (2.6) satisfies the conditions of the Picard–Lindelöf Theorem if the partial derivatives

$$\frac{\partial f}{\partial x_1}, \ldots, \frac{\partial f}{\partial x_n}$$

are continuous. The Picard–Lindelöf Theorem therefore guarantees the existence of a unique solution, at least locally, if we specify $\boldsymbol{x}(t_0)$. For the equation (2.6) this is equivalent to specifying

$$x(t_0), \dot{x}(t_0), \ddot{x}(t_0), \ldots, x^{(n-1)}(t_0),$$

i.e., the values of the function x and of its first $(n-1)$ derivatives at t_0.

Exercise 2.3 Show that it is impossible to find continuous functions $a_1, a_0 : \mathbb{R} \to \mathbb{R}$ so that the equation

$$\frac{d^2x}{dt^2}(t) + a_1(t)\frac{dx}{dt}(t) + a_0(t)x(t) = 0$$

has the solution $x(t) = t^2$.

Exercise 2.4 Consider a function $f : \mathbb{R} \times \mathbb{C} \to \mathbb{C}$ and the complex initial value problem

$$\frac{dz}{dt} = f(t, z), \quad z(0) = z_0 \tag{2.15}$$

for a complex valued function $z : \mathbb{R} \to \mathbb{C}$ and a complex number z_0. By considering real and imaginary parts of this equation formulate conditions for (2.15) to have a unique solution. Show that $z(t) = e^{it}$ and $z(t) = \cos t + i \sin t$ both satisfy

$$\frac{dz}{dt} = iz, \quad z(0) = 1,$$

and deduce Euler's formula $e^{it} = \cos t + i \sin t$.

Exercise 2.5 (i) Show that the real system

$$\dot{x} = ax - by, \quad \dot{y} = bx + ay$$

is equivalent to the single complex ODE

$$\dot{z} = cz, \tag{2.16}$$

where $z = x + iy$ and $c = a + ib$. Here a and b (and hence c) may be functions of x, y and t. Write z in modulus-argument form $z = re^{i\phi}$ (so that r, ϕ are polar coordinates in the (x, y)-plane) to show that (2.16) is, in turn, equivalent to

$$\dot{r} = ar, \qquad \dot{\phi} = b. \tag{2.17}$$

(ii) Use (i) to find the general solution of the two-dimensional system $\dot{\boldsymbol{x}} = \boldsymbol{f}(\boldsymbol{x})$, with $\boldsymbol{f}$ given in (ii) of Equation (2.12).
(iii) Also find the general solution of

$$\dot{x}(t) = x(t)\cos t - y(t)\sin t, \quad \dot{y} = x(t)\sin t + y(t)\cos t.$$

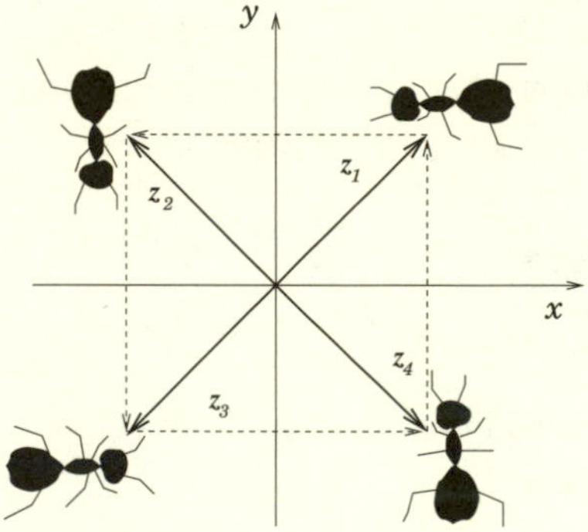

Fig. 2.1. Four ants chasing each other. How long will it take till they meet? Do Exercise 2.6 to find out

Exercise 2.6 Consider four ants at the four corners of a square, with edge length $\sqrt{2}$, as shown in Fig. 2.1. The ants chase each other: at all times each ant walks with constant speed v in the direction of its neighbour's current position (in an anti-clockwise direction). As all four ants move, their trajectories spiral towards the centre. The goal of this exercise is to compute the trajectories.

(i) We describe the ants' positions in terms of complex numbers z_1, z_2, z_3 and z_4, as shown in the diagram. Write down the differential equations governing the complex vector (z_1, z_2, z_3, z_4) combining the four ants' positions. Viewed as a real system, it should be a first order system of dimension eight, since every complex number is made up of two real numbers.
(ii) Show that, if (z_1, z_2, z_3, z_4) is a solution of the system found in (i), then so are (z_4, z_1, z_2, z_3) and (iz_1, iz_2, iz_3, iz_4).
(iii) Find the solution of the system with the initial condition

$$(z_1, z_2, z_3, z_4)(0) = \frac{1}{\sqrt{2}}(1+i, -1+i, -1-i, 1-i)$$

as follows. Use the Picard–Lindelöf Theorem and (ii) to show that any solution of the initial value problem has the property that, at all times t,

$$(z_1, z_2, z_3, z_4)(t) = (iz_4, iz_1, iz_2, iz_3)(t).$$

Deduce that the position $z_1(t)$ of the first ant as a function of time solves the initial value problem

$$\dot{z} = v\frac{(-1+i)}{\sqrt{2}}\frac{z}{r}, \quad z(0) = \frac{1}{\sqrt{2}}(1+i), \tag{2.18}$$

where $r = |z|$.
(iv) Solve the equation (2.18) for the first ant's trajectory by expressing it in polar coordinates r, ϕ, as in (2.17). Hence work out how long it takes for the ants to reach the centre of the square and find the number of times they have walked around the centre when they have reached it.
(v) Plot the first ant's trajectory. Check your answer by solving the initial value problem (2.18) numerically.

2.3 Linear systems

2.3.1 General remarks

In the remainder of this chapter we develop a powerful and general theory for linear systems, which are systems of the form

$$\frac{d\boldsymbol{x}}{dt}(t) = A(t)\boldsymbol{x}(t) + \boldsymbol{b}(t). \tag{2.19}$$

Here A is a $n \times n$ matrix-valued function and $\boldsymbol{b}$ is an $\mathbb{R}^n$-valued function of t. The theory we will develop also applies to n-th order linear ODEs of the form

$$x^{(n)}(t) + a_{n-1}(t)x^{(n-1)}(t) + \ldots + a_0(t)x(t) = b(t), \qquad (2.20)$$

for real-valued functions $a_0, \ldots, a_{n-1}$ and b, since they can be written in the form (2.19) with

$$A = \begin{pmatrix} 0 & 1 & \ldots & 0 & 0 \\ \vdots & & & & \vdots \\ 0 & 0 & \ldots & 0 & 1 \\ -a_0 & -a_1 & \ldots & -a_{n-2} & -a_{n-1} \end{pmatrix} \quad \text{and} \quad \boldsymbol{b} = \begin{pmatrix} 0 \\ 0 \\ \vdots \\ b \end{pmatrix}.$$

We use the notation $C^k(\mathbb{R}^m, \mathbb{R}^n)$ for k-times differentiable functions from $\mathbb{R}^m$ to $\mathbb{R}^n$ with a continuous k-th derivative. In terms of this notation, we assume that $\boldsymbol{b} \in C(\mathbb{R}, \mathbb{R}^n)$, $A \in C(\mathbb{R}, \mathbb{R}^{n^2})$ and we are looking for a solution $\boldsymbol{x} \in C^1(\mathbb{R}, \mathbb{R}^n)$ of (2.19). Comparing the right hand side of (2.19) with the right hand side of the general system (2.3) one checks that the derivative matrix appearing in the Picard–Lindelöf Theorem is

$$D\boldsymbol{f}(t, \boldsymbol{x}) = A(t)$$

and therefore continuous by the assumption we make on A. We therefore expect an initial value problem for an equation of the form (2.19) to have a unique solution in some interval containing the point t_0 where the initial condition is imposed. In fact, one can prove the following stronger result for linear systems, which assures the existence of a unique solution for all $t \in \mathbb{R}$ and not just in some interval. Again we refer the reader to Project 5.5 for a proof.

Theorem 2.2 *If $A : \mathbb{R} \to \mathbb{R}^{n^2}$ is a continuous matrix-valued function and $\boldsymbol{b} : \mathbb{R} \to \mathbb{R}^n$ is a continuous vector-valued function, then the linear differential equation* (2.19) *has a unique solution for any initial condition $\boldsymbol{x}(t_0) = \boldsymbol{x}_0$ and this solution exists for all $t \in \mathbb{R}$.*

Initial value problems for linear systems can be solved explicitly in special but rather important cases. In practice, one solves initial value problems in these cases by first finding the general solution and then imposing the initial condition. It also turns out that the general solution of (2.19) (recall that this is really the set of *all* solutions of (2.19), suitably parametrised by real constants) has interesting geometrical properties. For these reasons we study the general solution of (2.19) in detail in Sections 2.4 and 2.5. We do this using the language and methods of vector spaces and linear algebra, which we revise in the next section. First, however, we consider a simple example which motivates the tools we are going to use.

Example 2.3 Find the general solution of the linear system $\dot{x}_1 = x_2$, $\dot{x}_2 = x_1$.

Adding and subtracting the two equations, we deduce two first order ODEs

$$\frac{d}{dt}(x_1 + x_2) = (x_1 + x_2), \qquad \frac{d}{dt}(x_1 - x_2) = -(x_1 - x_2),$$

whose general solutions are

$$x_1(t) + x_2(t) = Ae^t, \quad x_1(t) - x_2(t) = Be^{-t}, \qquad A, B \in \mathbb{R}.$$

Solving for x_1, x_2, writing the result in vector notation and suitably re-defining the general constants we obtain the general solution

$$\begin{pmatrix} x_1(t) \\ x_2(t) \end{pmatrix} = \alpha \begin{pmatrix} \cosh t \\ \sinh t \end{pmatrix} + \beta \begin{pmatrix} \sinh t \\ \cosh t \end{pmatrix}, \qquad \alpha, \beta \in \mathbb{R}. \qquad \square$$

In the example, the general solution is a linear combination (or superposition) of the two solutions

$$\boldsymbol{y}^1(t) = \begin{pmatrix} \cosh t \\ \sinh t \end{pmatrix}, \qquad \boldsymbol{y}^2(t) = \begin{pmatrix} \sinh t \\ \cosh t \end{pmatrix}. \qquad (2.21)$$

In the language of linear algebra, the set of all solutions is therefore the span of $\{\boldsymbol{y}^1, \boldsymbol{y}^2\}$. As we shall see, this situation is typical for systems of the form (2.19) when $\boldsymbol{b}$ vanishes.

2.3.2 Linear algebra revisited

Recall that a vector space is a set V whose elements, called vectors, can be added together and multiplied by a number, often called a scalar. One also requires that V contains the zero vector, denoted 0. For our purposes, scalars may be real or complex numbers. A vector v in a vector space V is a *linear combination* of the vectors $v_1, \ldots, v_n$ if it can be written as

$$v = \alpha_1 v_1 + \ldots + \alpha_n v_n, \tag{2.22}$$

for $v_1, \ldots, v_n \in V$ and scalars $\alpha_1, \ldots, \alpha_n$. We say that a set $S = \{v_1, \ldots, v_n\}$ of vectors in V is *spanning* if any vector in V can be written as a linear combination of $v_1, \ldots, v_n$. Vectors $v_1, \ldots, v_n$ are called *linearly independent* if

$$\sum_{i=1}^{n} \alpha_i v_i = 0 \Rightarrow \alpha_i = 0 \text{ for } i = 1, \ldots, n. \tag{2.23}$$

A set $S = \{v_1, \ldots, v_n\}$ is called a *basis* of V if it is spanning and if its elements are linearly independent. Any vector $v \in V$ can be expressed as a linear combination (2.22) of the basis vectors in a unique fashion (prove this!). For any vector space there are infinitely many bases (still assuming that the scalars are real or complex numbers), but the number of elements in any basis is always the same; it is called the *dimension* of the vector space V and denoted $\dim(V)$.

If V, W are vector spaces, a linear map L is a map $L : V \to W$ such that $L(\alpha_1 v_1 + \alpha_2 v_2) = \alpha_1 L(v_1) + \alpha_2 L(v_2)$ for scalars α_1, α_2 and $v_1, v_2 \in V$. As for any map, we say that L is surjective if it is onto (for all $w \in W$ there is a $v \in V$ so that $L(v) = w$), we say that L is injective if it is one-to-one (if $L(v_1) = L(v_2)$ then $v_1 = v_2$) and we call L bijective if it is injective and surjective.

Exercise 2.7 If the review of vector spaces and linear maps in the previous paragraph was too quick, consult a book, the internet or a friend. Then prove the following results.
(i) If V, W are vector spaces and $L : V \to W$ is a linear map then the kernel of L, defined as $\ker L = \{\boldsymbol{v} \in V | L(\boldsymbol{v}) = 0\}$, is a vector space.
(ii) With the terminology of (i), show that L is injective if and only if $\ker L = \{0\}$.
(iii) Let $S = \{v_1, \ldots, v_n\}$ be a basis of the vector space V. Still continuing with the terminology of (i), show that, if L is bijec-

tive, the set $\{L(v_1), \ldots, L(v_n)\}$ is a basis of W. Deduce that dim(V)=dim(W) in this case.
(iv) For functions $\boldsymbol{f}_1, \boldsymbol{f}_2, \boldsymbol{f} \in C^1(\mathbb{R}, \mathbb{R}^n)$ we define an addition as $(\boldsymbol{f}_1 + \boldsymbol{f}_2)(t) = \boldsymbol{f}_1(t) + \boldsymbol{f}_2(t)$ and a multiplication by a real or complex number λ via $(\lambda \boldsymbol{f})(t) = \lambda \boldsymbol{f}(t)$. Show that, with the zero element represented by the function which is zero everywhere, $C^k(\mathbb{R}, \mathbb{R}^n)$ is a vector space for any $k \in \mathbb{Z}^{\geq 0}$, $n \in \mathbb{N}$.

To develop an intuition for the vector spaces $C^k(\mathbb{R}, \mathbb{R}^n)$, consider the case $k = 0$ and $n = 1$. Then the functions $x_m(t) = t^m$, with any non-negative integer power m, are elements of $C^0(\mathbb{R}, \mathbb{R})$. They are also linearly independent, as can be seen as follows. Any linear combination $p_I(t) = \sum_I a_i t^i$, where I is some finite subset of the non-negative integers, is a polynomial in t. Thus $p_I(t) = 0$ for all t implies that $a_i = 0$ for any $i \in I$ by the fundamental theorem of algebra. Therefore the functions x_m are linearly independent by the definition (2.23). Our argument can be extended to show that $C^k(\mathbb{R}, \mathbb{R}^n)$ contains infinitely many independent elements for any k and n. They are therefore examples of infinite-dimensional vector spaces.

With A defined as in Theorem 2.2 we now consider the map

$$L : C^1(\mathbb{R}, \mathbb{R}^n) \to C(\mathbb{R}, \mathbb{R}^n), \quad L[\boldsymbol{x}] = \frac{d\boldsymbol{x}}{dt} - A\boldsymbol{x}, \tag{2.24}$$

and observe that L is linear:

$$L[\alpha \boldsymbol{x} + \beta \boldsymbol{y}] = \alpha \frac{d\boldsymbol{x}}{dt} + \beta \frac{d\boldsymbol{y}}{dt} - \alpha A\boldsymbol{x} - \beta A\boldsymbol{y} = \alpha L[\boldsymbol{x}] + \beta L[\boldsymbol{y}].$$

The map L is an example of a *differential operator*. The word 'operator' is often used for linear maps defined on infinite-dimensional vector spaces. In our case, L acts on a certain space of functions. We have put square brackets around the argument of L to avoid confusion with the round brackets used for the argument of the functions to which L is applied. For example, if $n = 1$, $L = d/dt$ and $x(t) = t^2$ then we write $L[x](t) = 2t$.

Exercise 2.8 Let V be the space of all polynomials in one real variable t of degree at most 2. Show that the functions $1, t, t^2$ form a basis of V and give the matrix representation of $L = d/dt$ with respect to this basis.

2.4 Homogeneous linear systems

2.4.1 The vector space of solutions

With the notation of the previous section, the equation

$$L[\boldsymbol{x}] = \frac{d\boldsymbol{x}}{dt} - A\boldsymbol{x} = 0 \tag{2.25}$$

is called a homogeneous linear system. The space of solutions of (2.25) is characterised by the following theorem, which generalises the observations made in Example 2.3.

Theorem 2.3 (Solution space for homogeneous linear systems) *The space*

$$S = \{\boldsymbol{x} \in C^1(\mathbb{R}, \mathbb{R}^n) | L[\boldsymbol{x}] = 0\} \tag{2.26}$$

of solutions of the homogeneous equation (2.25) *is a vector space of dimension* n.

Our proof for this theorem uses the terminology and results from linear algebra developed in Section 2.3.2. If you find the proof difficult to follow on first reading, feel free to skip it and return to it after we have studied more examples of linear systems in this chapter and the next. However, make sure you absorb the message of Theorem 2.3: to find the general solution of a homogeneous linear system of n ODEs it is sufficient to find n independent solutions. All other solutions are simply linear combinations (superpositions) of these independent solutions - as illustrated by Example 2.3.

Proof The set S is a vector space since it is the kernel of the linear map L. The key idea for computing the dimension of S is to consider the evaluation map

$$\mathrm{Ev}_{t_0} : S \to \mathbb{R}^n, \quad \boldsymbol{x} \mapsto \boldsymbol{x}(t_0),$$

and to show that it is linear and bijective. We first note that Ev_{t_0} is a linear map:

$$\mathrm{Ev}_{t_0}[\alpha\boldsymbol{x} + \beta\boldsymbol{y}] = \alpha\boldsymbol{x}(t_0) + \beta\boldsymbol{y}(t_0) = \alpha\mathrm{Ev}_{t_0}[\boldsymbol{x}] + \beta\mathrm{Ev}_{t_0}[\boldsymbol{y}].$$

The Picard–Lindelöf Theorem implies that Ev_{t_0} is bijective: it is surjective (onto) since, for any given $\boldsymbol{x}_0 \in \mathbb{R}^n$, there exists solution $\boldsymbol{x} \in S$ so that $\boldsymbol{x}(t_0) = \boldsymbol{x}_0$; it is injective (one-to-one) since this solution is unique. Hence Ev_{t_0} is a bijective linear map. It follows (compare Exercise 2.7 (iii)) that $\dim(S) = \dim(\mathbb{R}^n) = n$. □

Exercise 2.9 Apply the evaluation map Ev_{t_0} to the solutions (2.21) for $t_0 = 0$ and confirm that you obtain a basis of $\mathbb{R}^2$.

Since a set of n independent solutions of (2.25) allows us to find the most general solution as a linear combination, we introduce a special name for such a set:

Definition 2.1 (Fundamental sets for homogeneous systems) A basis $\{\boldsymbol{y}^1, \ldots, \boldsymbol{y}^n\}$ of the set S of solutions defined in (2.26) is called a fundamental set of solutions of the linear, homogeneous system (2.25).

The following lemma provides a good way of checking if a given set of n solutions of (2.25) is independent. The lemma is an immediate consequence of Theorem 2.3 and the fact that n column vectors in $\mathbb{R}^n$ are linearly independent if and only if the $n \times n$ matrix obtained by writing the n column vectors next to each other has a non-zero determinant.

Lemma 2.1 (Determinant for fundamental sets) *A set of n solutions $\boldsymbol{y}^1, \ldots, \boldsymbol{y}^n$ of the homogeneous linear equation* (2.25) *forms a fundamental set of solutions if and only if the matrix*

$$Y = \begin{pmatrix} y_1^1 & \cdots & y_1^n \\ \vdots & & \vdots \\ y_n^1 & \cdots & y_n^n \end{pmatrix} \tag{2.27}$$

has a non-zero determinant det(Y) at one value $t = t_0$ (and hence for all values of t).

It follows from our discussion of (2.20) that homogeneous linear n-th order equations

$$x^{(n)} + a_{n-1}x^{(n-1)} + \ldots + a_0 x = 0 \tag{2.28}$$

are special cases of linear homogeneous systems (2.25). Here we define

$$s = \{x \in C^n(\mathbb{R}, \mathbb{R}) | x^{(n)} + a_{n-1}x^{(n-1)} + \ldots + a_0 x = 0\}.$$

The evaluation map takes the form

$$\mathrm{ev}_{t_0} : s \to \mathbb{R}^n, \quad x \mapsto \begin{pmatrix} x(t_0) \\ \dot{x}(t_0) \\ \vdots \\ x^{(n-1)}(t_0) \end{pmatrix} \tag{2.29}$$

and assigns to each solution the vector made out of the values of the function x and its first $(n-1)$ derivatives at t_0. Theorem 2.3 then states that this map is linear and bijective. This is sometimes useful for checking linear independence of solutions of an n-th order homogeneous linear equation: two solutions are independent if and only if their images under ev_{t_0} are independent in $\mathbb{R}^n$. Applied to the two solutions $y_1(t) = \cos t$ and $y_2(t) = \sin t$ of $\ddot{x} + x = 0$, for example, the evaluation map at $t_0 = 0$ gives the two independent vectors

$$\mathrm{ev}_0(y_1) = \begin{pmatrix} 1 \\ 0 \end{pmatrix}, \qquad \mathrm{ev}_0(y_2) = \begin{pmatrix} 0 \\ 1 \end{pmatrix}.$$

The counterpart of Lemma 2.1 has a special name for n-th order equations:

Lemma 2.2 (Wronskian for n-th order linear homogeneous equations) *A set of n solutions $u_1, \ldots, u_n$ of the n-th order homogeneous linear equation* (2.28) *forms a fundamental set of solutions if and only if the matrix*

$$U = \begin{pmatrix} u_1 & \ldots & u_n \\ \dot{u}_1 & \ldots & \dot{u}_n \\ \vdots & & \vdots \\ u_1^{(n-1)} & \ldots & u_n^{(n-1)} \end{pmatrix}$$

has a non-zero determinant W=det(U) at one value $t = t_0$ (and hence for all values of t). The determinant W of U is called the Wronskian of $u_1, \ldots, u_n$.

Example 2.4 The Wronskian of two solutions u_1, u_2 of the second order equation $\ddot{x} + a_1\dot{x} + a_0 x = 0$ is the function

$$W[u_1, u_2](t) = u_1(t)\dot{u}_2(t) - u_2(t)\dot{u}_1(t). \tag{2.30}$$

Exercise 2.10 (Abel's Formula) Suppose two functions u_1, u_2 satisfy the homogeneous linear ODE $\ddot{x} + a_1\dot{x} + a_0 x = 0$, where a_1, a_0 are continuous functions on $\mathbb{R}$. Derive a differential equation for the Wronskian (2.30) and deduce that it can be expressed as

$$W[u_1, u_2](t) = C\exp(-\int_{t_0}^{t} a_1(\tau)d\tau),$$

where C, t_0 are real constants. Given that $x(t) = e^{-t}$ solves the ODE $\ddot{x} + 2\dot{x} + x = 0$, use this result to find a second, independent solution.

2.4.2 The eigenvector method

In practice, a fundamental set of solutions of the homogeneous system (2.25) can only be found explicitly in rare, lucky cases. An example is the case where the matrix A is constant, which we study in this section. Consider the system of linear equations

$$\dot{\boldsymbol{x}} = A\boldsymbol{x}, \tag{2.31}$$

where $\boldsymbol{x} : \mathbb{R} \to \mathbb{R}^n$ and A is a constant, real $n \times n$ matrix. To find solutions we make the *ansatz* (=educated guess)

$$\boldsymbol{x}(t) = f(t)\boldsymbol{v}, \tag{2.32}$$

where $\boldsymbol{v}$ is a time-independent vector. This ansatz is an analogue of the trick of separating variables when solving a partial differential equation: we have assumed that the time dependence is the same for all components of the vector-valued function $\boldsymbol{x}$, i.e., we have separated the time dependence from the vectorial nature of the solution. Inserting (2.32) into (2.31) we find

$$\frac{\dot{f}(t)}{f(t)}\boldsymbol{v} = A\boldsymbol{v},$$

where we assumed $f(t) \neq 0$. The right hand side of this equation is independent of t, so that we deduce $\dot{f} = \lambda f$ for some constant λ, as well as

$$A\boldsymbol{v} = \lambda\boldsymbol{v}.$$

Hence, (2.32) is a solution of (2.31) if and only if $\boldsymbol{v}$ is an eigenvector of A and $f(t) = e^{\lambda t}$ (we can absorb the arbitrary constant into a re-definition of the eigenvector $\boldsymbol{v}$). Since all $n \times n$ matrices have at least one eigenvalue and one eigenvector (both possibly complex), we have the solution

$$\boldsymbol{x}(t) = e^{\lambda t}\boldsymbol{v}.$$

If A has n independent eigenvectors $\boldsymbol{v}^1, \ldots, \boldsymbol{v}^n$ with real eigenvalues $\lambda_1, \ldots, \lambda_n$ (which need not be distinct), a fundamental set of solutions of (2.31) is given by

$$\boldsymbol{y}^1(t) = e^{\lambda_1 t}\boldsymbol{v}^1, \ldots, \boldsymbol{y}^n(t) = e^{\lambda_n t}\boldsymbol{v}^n.$$

Unfortunately, not all real $n \times n$ matrices have an associated basis of eigenvectors with real eigenvalues. To understand the different situations which can arise, it is sufficient to consider 2×2 matrices.

(i) **Distinct real eigenvalues**

Consider the equation (2.31) for $n = 2$ and

$$A = \begin{pmatrix} 1 & 4 \\ 1 & 1 \end{pmatrix}.$$

This matrix has eigenvalue λ if $\det(A - \lambda I) = (1-\lambda)^2 - 4 = 0$, i.e., if $\lambda = -1$ or $\lambda = 3$. The vector $\boldsymbol{v} = \begin{pmatrix} v_1 \\ v_2 \end{pmatrix}$ is an eigenvector for the eigenvalue -1 if

$$v_1 + 4v_2 = -v_1, \quad \text{and} \quad v_1 + v_2 = -v_2.$$

Hence $\boldsymbol{v}^1 = \begin{pmatrix} 2 \\ -1 \end{pmatrix}$ is an eigenvector for the eigenvalue -1. Similarly, one computes that $\boldsymbol{v}^2 = \begin{pmatrix} 2 \\ 1 \end{pmatrix}$ is an eigenvector for the eigenvalue 3. The solutions

$$\boldsymbol{y}^1(t) = e^{-t}\begin{pmatrix} 2 \\ -1 \end{pmatrix}, \qquad \boldsymbol{y}^2(t) = e^{3t}\begin{pmatrix} 2 \\ 1 \end{pmatrix}$$

are independent and form a fundamental set.

(ii) **Complex eigenvalues**

If the real matrix A in (2.31) has a complex eigenvalue $\lambda = \alpha + i\beta$ with corresponding eigenvector $\boldsymbol{v} = \boldsymbol{v}^1 + i\boldsymbol{v}^2$, then $\boldsymbol{y}(t) = e^{\lambda t}\boldsymbol{v}$ is a solution, and we obtain two real solutions by taking the real and imaginary parts. You are asked to justify this procedure in Exercise 2.11.

As an example, consider the equation (2.31) with

$$A = \begin{pmatrix} -1 & -5 \\ 1 & 3 \end{pmatrix}.$$

The eigenvalues are $1 \pm i$, with eigenvectors $\begin{pmatrix} -2 \pm i \\ 1 \end{pmatrix}$. Thus we have the complex solution

$$\begin{aligned} \boldsymbol{y}(t) &= e^{(1+i)t} \begin{pmatrix} -2+i \\ 1 \end{pmatrix} \\ &= e^t(\cos t + i \sin t) \begin{pmatrix} -2+i \\ 1 \end{pmatrix} \\ &= e^t \begin{pmatrix} -2\cos t - \sin t \\ \cos t \end{pmatrix} + ie^t \begin{pmatrix} \cos t - 2\sin t \\ \sin t \end{pmatrix}, \end{aligned}$$

so that a fundamental set of solutions is

$$\boldsymbol{y}^1(t) = e^t \begin{pmatrix} -2\cos t - \sin t \\ \cos t \end{pmatrix}, \quad \boldsymbol{y}^2(t) = e^t \begin{pmatrix} \cos t - 2\sin t \\ \sin t \end{pmatrix}.$$

(iii) **Repeated eigenvalues, distinct eigenvectors**

If a 2×2 matrix has one eigenvalue with two independent eigenvectors, it is necessarily diagonal. For example, the matrix

$$A = \begin{pmatrix} 2 & 0 \\ 0 & 2 \end{pmatrix}$$

has $\boldsymbol{v}^1 = \begin{pmatrix} 1 \\ 0 \end{pmatrix}$ and $\boldsymbol{v}^2 = \begin{pmatrix} 0 \\ 1 \end{pmatrix}$ as eigenvectors for the repeated eigenvalue 2. We find a fundamental set as for case (i):

$$\boldsymbol{y}^1(t) = e^{2t} \begin{pmatrix} 1 \\ 0 \end{pmatrix}, \quad \boldsymbol{y}^2(t) = e^{2t} \begin{pmatrix} 0 \\ 1 \end{pmatrix}.$$

(iv) **Repeated eigenvalues, one eigenvector**

A matrix may fail to have a basis of eigenvectors if an eigenvalue is repeated. Suppose that a real eigenvalue λ_1 has multiplicity 2 (i.e., the characteristic polynomial contains a factor $(\lambda - \lambda_1)^2$) and that $\boldsymbol{v}$ is the only eigenvector for this eigenvalue. Then $A\boldsymbol{v} = \lambda_1 \boldsymbol{v}$ and $\boldsymbol{y}^1(t) = e^{\lambda_1 t}\boldsymbol{v}$ is a solution of (2.31). To find a second solution we try

$$\boldsymbol{y}^2(t) = te^{\lambda_1 t}\boldsymbol{v} + e^{\lambda_1 t}\boldsymbol{w}. \tag{2.33}$$

(One can motivate this guess by starting with case (i) and taking the limit where the eigenvalues coincide, cf. also Exercise 3.3.) Inserting into (2.31), we find that this is a solution if

$$e^{\lambda_1 t}(\lambda_1 t\boldsymbol{v} + \boldsymbol{v} + \lambda_1 \boldsymbol{w}) = e^{\lambda_1 t} A(t\boldsymbol{v} + \boldsymbol{w}),$$

i.e., if

$$(A - \lambda_1 I)\boldsymbol{w} = \boldsymbol{v}.$$

One can show that this equation can always be solved for $\boldsymbol{w}$. Since $(A - \lambda_1 I)\boldsymbol{w} \neq 0$ but $(A - \lambda_1 I)^2\boldsymbol{w} = 0$, $\boldsymbol{w}$ is sometimes called a generalised eigenvector of A.

As an example, consider the equation (2.31) with

$$A = \begin{pmatrix} 1 & 9 \\ -1 & -5 \end{pmatrix}.$$

We find that λ is an eigenvalue if $(\lambda + 2)^2 = 0$. Hence -2 is a repeated eigenvalue. The only eigenvector is $\boldsymbol{v} = \begin{pmatrix} -3 \\ 1 \end{pmatrix}$, so that one solution of (2.31) is

$$\boldsymbol{y}^1(t) = e^{-2t} \begin{pmatrix} -3 \\ 1 \end{pmatrix}.$$

To find the second solution we need to solve

$$(A + 2I)\boldsymbol{w} = \boldsymbol{v} \Leftrightarrow \begin{pmatrix} 3 & 9 \\ -1 & -3 \end{pmatrix} \boldsymbol{w} = \begin{pmatrix} -3 \\ 1 \end{pmatrix}. \tag{2.34}$$

We observe that $\boldsymbol{w}$ is not uniquely determined. This is fine - we simply pick a solution of the equation (2.34). With

$$\boldsymbol{w} = \begin{pmatrix} -1 \\ 0 \end{pmatrix},$$

a second solution is given by

$$\boldsymbol{y}^2(t) = e^{-2t} t \begin{pmatrix} -3 \\ 1 \end{pmatrix} + e^{-2t} \begin{pmatrix} -1 \\ 0 \end{pmatrix} = e^{-2t} \begin{pmatrix} -1 - 3t \\ t \end{pmatrix}.$$

Exercise 2.11 Show that, if $\boldsymbol{x}(t)$ is a complex solution of (2.31), then the real and imaginary parts of $\boldsymbol{x}(t)$ are both real solutions of (2.31).

Exercise 2.12 Find a fundamental set of solutions for each of the following two-dimensional systems.

$$\text{(i)}\ \dot{\boldsymbol{x}} = \begin{pmatrix} 1 & -5 \\ 1 & -3 \end{pmatrix} \boldsymbol{x}, \qquad \text{(ii)}\ \dot{\boldsymbol{x}} = \begin{pmatrix} 3 & 2 \\ -2 & -1 \end{pmatrix} \boldsymbol{x}.$$

Exercise 2.13 By trying solutions of the form $x(t) = e^{\lambda t}$ find the general solution of the third order equation

$$\frac{d^3x}{dt^3} - \frac{d^2x}{dt^2} - 4\frac{dx}{dt} + 4x = 0.$$

Note: this method for an n-th order equation is much quicker than writing the equation as a first order system and applying the eigenvector method. If you are not convinced, try both methods and compare!

2.5 Inhomogeneous linear systems

We return to the more general form (2.19) of a linear system with $\boldsymbol{b} \neq 0$. Such systems are called *inhomogeneous* linear systems. Using again the linear operator L defined in (2.24) we have the following analogue of Theorem 2.3.

Theorem 2.4 (Solution space for inhomogeneous linear systems) *Let $\{\boldsymbol{y}^1, \dots, \boldsymbol{y}^n\}$ be a fundamental set of solutions for the homogeneous equation $\dot{\boldsymbol{x}} = A\boldsymbol{x}$. Then any solution of the inhomogeneous equation* (2.19) *is of the form*

$$\boldsymbol{x}(t) = \sum_{i=1}^{n} c_i \boldsymbol{y}^i(t) + \boldsymbol{x}_p(t), \tag{2.35}$$

where $\boldsymbol{x}_p$ is a particular solution of (2.19) *and $c_1, \dots, c_n$ are real constants.*

Proof To check that (2.35) satisfies the inhomogeneous equation (2.19) we use the fact that $\dot{\boldsymbol{y}}^i = A\boldsymbol{y}^i$ and $\dot{\boldsymbol{x}}_p = A\boldsymbol{x}_p + \boldsymbol{b}$, and find

$$\dot{\boldsymbol{x}} = \sum_{i=1}^{n} c_i A\boldsymbol{y}^i(t) + A\boldsymbol{x}_p(t) + \boldsymbol{b} = A\boldsymbol{x} + \boldsymbol{b},$$

as required. To show that *every* solution can be written in this way, suppose that $\boldsymbol{x}$ is a solution of (2.19). Then $\boldsymbol{x} - \boldsymbol{x}_p$ satisfies the homogeneous equation $\frac{d}{dt}(\boldsymbol{x} - \boldsymbol{x}_p) = A(\boldsymbol{x} - \boldsymbol{x}_p)$ and therefore can be expanded in the form

$$\boldsymbol{x} - \boldsymbol{x}_p = \sum_{i=1}^{n} c_i \boldsymbol{y}^i \tag{2.36}$$

for some real constants $c_1, \ldots, c_n$. □

As in the homogeneous case, our general result (2.35) for inhomogeneous systems has an immediate corollary for n-th order equations. In order to find the general solution of an inhomogeneous, linear equation of the form (2.20), we need a fundamental set $\{y_1, \ldots, y_n\}$ of solutions for the associated homogeneous equation (2.28) and a particular solution x_p of (2.20). The general solution of (2.20) is then given by

$$x(t) = \sum_{i=1}^{n} c_i y_i(t) + x_p(t).$$

Exercise 2.14 Three solutions of

$$y''(x) + a_1(x)y'(x) + a_0(x)y(x) = f(x) \tag{2.37}$$

are $u_1(x) = x,\quad u_2(x) = x + e^x$ and $u_3(x) = 1 + x + e^x$. Determine the solution of equation (2.37) satisfying the initial conditions $y(0) = 1$ and $y'(0) = 3$. *Hint*: use Theorem 2.4.

Exercise 2.15 Look up the definition of an *affine space*. Show that the set of solutions of an inhomogeneous linear system, as described in Theorem 2.4, is an affine space.

The most systematic way of finding a particular solution of an inhomogeneous linear equation is the method of *variation of the parameters*. The idea is to look for a particular solution of the form

$$\boldsymbol{x}_p(t) = \sum_{i=1}^{n} c_i(t) \boldsymbol{y}^i(t), \tag{2.38}$$

where $\{\boldsymbol{y}^1, \ldots, \boldsymbol{y}^n\}$ is a fundamental set of the homogeneous equations as before, but the c_i are now functions of t. In terms of the matrix Y defined in (2.27) and the vector-valued function

$$\boldsymbol{c} = \begin{pmatrix} c_1 \\ \vdots \\ c_n \end{pmatrix}$$

we can write (2.38) as

$$\boldsymbol{x}_p = Y\boldsymbol{c}.$$

Using the product rule and the fact that each of the $\boldsymbol{y}^i$ making up the matrix Y satisfy the homogeneous equation we deduce

$$\dot{\boldsymbol{x}}_p = \dot{Y}\boldsymbol{c} + Y\dot{\boldsymbol{c}} = AY\boldsymbol{c} + Y\dot{\boldsymbol{c}}. \tag{2.39}$$

Hence

$$\begin{aligned} \dot{\boldsymbol{x}}_p = A\boldsymbol{x}_p + \boldsymbol{b} \quad &\Leftrightarrow \quad AY\boldsymbol{c} + Y\dot{\boldsymbol{c}} = AY\boldsymbol{c} + \boldsymbol{b} \\ &\Leftrightarrow \quad \dot{\boldsymbol{c}} = Y^{-1}\boldsymbol{b}. \end{aligned}$$

Note that Y is invertible by virtue of $\boldsymbol{y}^1, \ldots, \boldsymbol{y}^n$ being a fundamental set of solutions. We can now compute $\boldsymbol{c}$ by integration (at least in principle) and thus obtain a particular solution. We summarise this result as follows.

Theorem 2.5 (Method of variation of the parameters) *Let $\{\boldsymbol{y}^1, \ldots, \boldsymbol{y}^n\}$ be a fundamental set of solutions of $\dot{\boldsymbol{x}} = A\boldsymbol{x}$ and let Y be the matrix constructed from $\{\boldsymbol{y}^1, \ldots, \boldsymbol{y}^n\}$ according to* (2.27)*. Then*

$$\boldsymbol{x}_p(t) = Y(t)\int_{t_0}^{t} Y^{-1}(\tau)\boldsymbol{b}(\tau)d\tau \tag{2.40}$$

is a particular solution of the inhomogeneous equation $\dot{\boldsymbol{x}} = A\boldsymbol{x}+\boldsymbol{b}$.

The formula (2.40) is very general, and therefore very powerful. To get a feeling for how it works you need to study examples, which are provided in the exercises below. You may also find it useful to write out the steps leading from (2.38) to (2.40) for $n = 2$, spelling out the various matrix multiplications.

Note that for $n = 1$ the formula (2.40) reduces to the formula for solutions of linear first order equations (1.9) based on the integrating factor (1.10). To see this we consider a first order linear

ODE in the form

$$\dot{x}(t) = a(t)x(t) + b(t) \Leftrightarrow \dot{x}(t) - a(t)x(t) = b(t) \tag{2.41}$$

and note that the homogeneous equation $\dot{x}(t) = a(t)x(t)$ has the solution $y(t) = \exp(\int_{t_0}^t a(\tau)d\tau)$, which is the inverse of the integrating factor $I(t) = \exp(-\int_{t_0}^t a(\tau)d\tau)$ for (2.41). Thus the formula

$$x_p(t) = y(t)\int_{t_0}^t y^{-1}(\tau)b(\tau)d\tau$$

for the particular solution is precisely the solution we would obtain using the integrating factor $y^{-1}(t)$.

Exercise 2.16 Solve the following initial value problem

$$\dot{\boldsymbol{x}}(t) = \frac{1}{2}\begin{pmatrix} 3 & 1 & 0 \\ -1 & 5 & 0 \\ 0 & 0 & 6 \end{pmatrix}\boldsymbol{x}(t) + \begin{pmatrix} t \\ 0 \\ t^2 \end{pmatrix}, \quad \boldsymbol{x}(0) = \begin{pmatrix} 0 \\ 2 \\ 1 \end{pmatrix}.$$

Exercise 2.17 (Variation of the parameters for second order ODEs) Write the second order inhomogeneous linear differential equation

$$\ddot{x}(t) + a_1(t)\dot{x}(t) + a_0(t)x(t) = f(t) \tag{2.42}$$

as a two-dimensional system $\dot{\boldsymbol{x}} = A\boldsymbol{x} + \boldsymbol{b}$ where $\boldsymbol{x} = \begin{pmatrix} x_1 \\ x_2 \end{pmatrix}$, and A and $\boldsymbol{b}$ are to be determined. Let u_1 and u_2 be a fundamental set of solutions for the homogeneous equation $\ddot{x}(t) + a_1(t)\dot{x}(t) + a_0(t)x(t) = 0$. By using the method of variation of parameters show that

$$x_p(t) = \int_0^t \frac{u_2(t)u_1(\tau) - u_1(t)u_2(\tau)}{W[u_1,u_2](\tau)} f(\tau)\,d\tau, \tag{2.43}$$

with the Wronskian W defined in (2.30), is a solution of (2.42). Hence show that

$$\ddot{x}(t) + x(t) = f(t)$$

has

$$x_p(t) = \int_0^t \sin(t-\tau)f(\tau)\,d\tau$$

as a particular solution.

3

Second order equations and oscillations

Differential equations of second order occur frequently in applied mathematics, particularly in applications coming from physics and engineering. The main reason for this is that most fundamental equations of physics, like Newton's second law of motion (2.7), are second order differential equations. It is not clear why Nature, at a fundamental level, should obey second order differential equations, but that is what centuries of research have taught us.

Since second order ODEs and even systems of second order ODEs are special cases of systems of first order ODEs, one might think that the study of second order ODEs is a simple application of the theory studied in Chapter 2. This is true as far as general existence and uniqueness questions are concerned, but there are a number of elementary techniques especially designed for solving second order ODEs which are more efficient than the general machinery developed for systems. In this chapter we briefly review these techniques and then look in detail at the application of second order ODEs in the study of oscillations. If there is one topic in physics that every mathematician should know about then this is it. Much understanding of a surprisingly wide range of physical phenomena can be gained from studying the equation governing a mass attached to a spring and acted on by an external force. Conversely, one can develop valuable intuition about second order differential equations from a thorough understanding of the physics of the oscillating spring.

3.1 Second order differential equations

3.1.1 *Linear, homogeneous ODEs with constant coefficients*

The general solution of the homogeneous second order ODE

$$\frac{d^2x}{dt^2} + a_1\frac{dx}{dt} + a_0 x = 0, \qquad a_0, a_1 \in \mathbb{R}, \tag{3.1}$$

could, in principle, be found by writing the equation as a first order system and using the eigenvector method of the previous chapter. However, in practice it is quicker to find the general solution by an elementary method which we summarise schematically, leaving some details to the exercises below.

(i) Look for a solution of the form $x(t) = e^{\lambda t}$, where λ is a real or complex constant to be determined.

(ii) Insert $x(t) = e^{\lambda t}$ into (3.1) to deduce the *characteristic equation* for λ:

$$\lambda^2 + a_1\lambda + a_0 = 0. \tag{3.2}$$

(iii) Depending on the nature of the roots

$$\lambda_1 = \frac{-a_1 + \sqrt{a_1^2 - 4a_0}}{2}, \qquad \lambda_2 = \frac{-a_1 - \sqrt{a_1^2 - 4a_0}}{2},$$

of the characteristic equation, obtain the fundamental set of solutions of (3.1) according to Table 3.1; the general solution of (3.1) is $x(t) = Ax_1(t) + Bx_2(t)$ for arbitrary constants A, B.

Exercise 3.1 Use the evaluation map (2.29) at $t_0 = 0$ to check that, in each case, the solutions in the fundamental sets listed in Table 3.1 are linearly independent.

Exercise 3.2 Consider the case where the characteristic equation (3.2) has complex roots. Explain why the roots are each other's complex conjugate in that case, and check that the fundamental set of solutions given in Table 3.1 for the case are the real and imaginary parts of $x(t) = e^{\lambda t}$. Explain why the real and imaginary parts of a solution are again solutions (compare Exercise 2.11).

Table 3.1. *Fundamental sets of solutions for equation* (3.1)

Roots	Fundamental set of solutions
λ_1, λ_2 real, distinct	$x_1(t) = e^{\lambda_1 t}, \quad x_2(t) = e^{\lambda_2 t}$
$\lambda_1 = \lambda_2$ real	$x_1(t) = e^{\lambda_1 t}, \quad x_2(t) = te^{\lambda_1 t}$
$\lambda_1 = \alpha + i\beta, \lambda_2 = \alpha - i\beta$	$x_1(t) = e^{\alpha t}\cos(\beta t),\ x_2(t) = e^{\alpha t}\sin(\beta t)$

Exercise 3.3 The goal of this exercise is to show that the fundamental set of solutions given in Table 3.1 in the case where the characteristic equation has one real root arises as a limit of the case of distinct roots. We consider the second order ODE (3.1) and assume that the roots λ_1 and λ_2 of the characteristic equation are real and distinct.
(i) Show, using the method of Exercise 3.1, that

$$u_1(t) = \frac{1}{2}(e^{\lambda_1 t} + e^{\lambda_2 t}), \quad u_2(t) = \frac{1}{\lambda_1 - \lambda_2}(e^{\lambda_1 t} - e^{\lambda_2 t})$$

form a fundamental set of solutions.
(ii) Now take the limit $\lambda_2 \to \lambda_1$ for both solutions and check that you obtain the fundamental set given in Table 3.1 for the case $\lambda_1 = \lambda_2$.

Exercise 3.4 Find the general solution of the following ODEs ($\dot{x} = \frac{dx}{dt}, \ddot{x} = \frac{d^2x}{dt^2}$): (i) $\ddot{x} + 4\dot{x} + 3x = 0$,
(ii) $\ddot{x} + 4\dot{x} + 4x = 0$, (iii) $\ddot{x} - 2\dot{x} + 2x = 0$.

Exercise 3.5 If a, b, c are positive constants prove that every solution of the ODE $a\ddot{x} + b\dot{x} + cx = 0$ must tend to 0 as $t \to \infty$.

3.1.2 Inhomogeneous linear equations

We now consider the inhomogeneous equation

$$\frac{d^2x}{dt^2} + a_1\frac{dx}{dt} + a_0 x = f, \tag{3.3}$$

where f is some continuous function of t. From Section 2.5 we have the following recipe for finding general solutions:

(i) Find a fundamental set $\{x_1, x_2\}$ of solutions of the homogeneous equation $\frac{d^2x}{dt^2} + a_1\frac{dx}{dt} + a_0x = 0$.
(ii) Find a particular solution x_p of (3.3).
(iii) Obtain the general solution $x = Ax_1 + Bx_2 + x_p$ of (3.3).

A particular solution can, in principle, always be found by applying the method of variation of the parameters, as applied to equations of the form (3.3) in Exercise 2.17. However, in practice it is sometimes quicker to find a particular solution with the help of an educated guess. The method of educated guessing is called the *method of undetermined coefficients* in this context, and works when the function f in (3.3) is of a simple form, e.g., if it is a polynomial, or an exponential, or a trigonometric function. The basic idea is to try a function for x_p which is such that a linear combination of x_p with its first and second derivative can conceivably produce the function f. Table 3.2 gives recipes involving unknown coefficients which one can determine by substituting into the equation. There is no deep reason for these recipes other than that they work.

Example 3.1 Find a particular solution of $\ddot{x} + x = t^2$.

We try $x_p(t) = b_0 + b_1t + b_2t^2$, and find, by inserting into the differential equation, that $(2b_2 + b_0) + b_1t + b_2t^2 = t^2$. Comparing coefficients yields $b_2 = 1$ $b_1 = 0$, and $b_0 = -2$, so $x_p(t) = t^2 - 2$. □

Example 3.2 Find particular solutions of

$$\text{(i)}\ \ddot{x} + 3\dot{x} + 2x = e^{3t} \quad \text{(ii)}\ \ddot{x} + 3\dot{x} + 2x = e^{-t}.$$

The characteristic equation of the homogeneous equation $\ddot{x}+3\dot{x}+2x = 0$ has roots -1 and -2. Thus, for (i) the right hand side is not a solution of the homogeneous equation, and we try $x_p(t) = ce^{3t}$. Inserting into the equation (i) we deduce that x_p is a solution provided we choose $c = 1/20$. For (ii), we need to try $x_p(t) = cte^{-t}$. After slightly tedious differentiations, we find that this is indeed a solution provided we pick $c = 1$. □

Table 3.2. *The method of undetermined coefficients for* (3.3); *HE stands for 'homogeneous equation'*

$f(t)$	Particular solution
$b_0 + b_1 t + ... b_n t^n$	$c_0 + c_1 t + ... c_n t^n$
$e^{\lambda t}$	if $e^{\lambda t}$ does not solve HE, try $ce^{\lambda t}$, $c \in \mathbb{R}$ if $e^{\lambda t}$ solves HE, try $cte^{\lambda t}$, $c \in \mathbb{R}$ if $e^{\lambda t}$ and $te^{\lambda t}$ solve HE, try $ct^2e^{\lambda t}$, $c \in \mathbb{R}$
$\cos(\omega t)$	if $e^{i\omega t}$ does not solve HE, try $Ce^{i\omega t}$, $C \in \mathbb{C}$, and take real part if $e^{i\omega t}$ solves HE, try $Cte^{i\omega t}$, $C \in \mathbb{C}$, and take real part
$\sin(\omega t)$	if $e^{i\omega t}$ does not solve HE, try $Ce^{i\omega t}$, $C \in \mathbb{C}$, and take imaginary part if $e^{i\omega t}$ solves HE, try $Cte^{i\omega t}$, $C \in \mathbb{C}$, and take imaginary part

Finally, the oscillatory case $f(t) = \cos\omega t$ is particularly important for applications; we will study it in detail in Section 3.2.

Exercise 3.6 Use the formula (2.43) from the method of variation of the parameters to find a particular solution of the ODE given under (ii) in Example 3.2. Is your result consistent with our findings in Example 3.2?

3.1.3 Euler equations

Differential equations of the form

$$x^2 \frac{d^2y}{dx^2} + a_1 x \frac{dy}{dx} + a_0 y = 0, \tag{3.4}$$

where a_0 and a_1 are constants, are named after the Swiss mathematician Leonhard Euler (1707–1783). Such equations can be transformed into a linear ODE with constant coefficients by writing them in terms of the independent variable $t = \ln|x|$. Then

$$\frac{dy}{dx} = \frac{dy}{dt}\frac{dt}{dx} = \frac{1}{x}\frac{dy}{dt} \qquad \text{and}$$

$$\frac{d^2y}{dx^2} = \frac{d}{dx}\left(\frac{1}{x}\frac{dy}{dt}\right) = -\frac{1}{x^2}\frac{dy}{dt} + \frac{1}{x}\frac{dt}{dx}\frac{d^2y}{dt^2} = -\frac{1}{x^2}\frac{dy}{dt} + \frac{1}{x^2}\frac{d^2y}{dt^2}$$

Hence the equation (3.4) becomes

$$\frac{d^2y}{dt^2} + (a_1 - 1)\frac{dy}{dt} + a_0 y = 0,$$

which has constant coefficients and can be solved with the methods of Section 3.1.1.

Example 3.3 Find a fundamental set of solutions of

$$x^2\frac{d^2y}{dx^2} + 2x\frac{dy}{dx} + y = 0. \tag{3.5}$$

In terms of the variable $t = \ln|x|$ the equation becomes

$$\frac{d^2y}{dt^2} + \frac{dy}{dt} + y = 0. \tag{3.6}$$

The characteristic equation $\lambda^2 + \lambda + 1 = 0$ has the roots $\lambda_1 = -\frac{1}{2} + \frac{\sqrt{3}}{2}i$ and $\lambda_2 = -\frac{1}{2} - \frac{\sqrt{3}}{2}i$. Hence a fundamental set of solutions of (3.6) is

$$y_1(t) = e^{-\frac{t}{2}}\cos(\frac{\sqrt{3}}{2}t), \qquad y_2(t) = e^{-\frac{t}{2}}\sin(\frac{\sqrt{3}}{2}t).$$

Transforming back to x we obtain

$$y_1(x) = x^{-\frac{1}{2}}\cos(\frac{\sqrt{3}}{2}\ln|x|), \qquad y_2(x) = x^{-\frac{1}{2}}\sin(\frac{\sqrt{3}}{2}\ln|x|),$$

as a fundamental set of solutions for (3.5). □

The Euler equation (3.4) does not satisfy the conditions of the Picard–Lindelöf Theorem at $x = 0$. For the equation (3.5) we found solutions which are indeed singular at $x = 0$. However, you should be able to check that the Euler equation

$$x^2\frac{d^2y}{dx^2} - 2x\frac{dy}{dx} + 2y = 0$$

has the general solution $y(x) = Ax + Bx^2$ which is smooth at $x = 0$. The general investigation of linear ODEs with singularities like the ones in the Euler equation is a very interesting and important area which, unfortunately, we cannot study in this book for lack of space. Interestingly, it turns out that singularities are best studied by allowing the *argument* x (and not just the function y) to be complex. If you want to read up about this at an introductory level you should look for books that discuss *series solutions* or the *Frobenius method* for solving linear ODEs.

Exercise 3.7 One can also solve the Euler equation (3.4) more directly by looking for solutions of the form $y(x) = x^\lambda$. Re-derive the fundamental set of solutions of Example (3.5) by this method.

Exercise 3.8 Use any of the methods in this section, or the method of variation of the parameters, to find general solutions of the following linear second order ODEs.

(i) $\frac{d^2y}{dx^2}(x) + y(x) = \sec(x)$,
(ii) $\ddot{x}(t) - 4\dot{x}(t) + 3x(t) = 2e^t$,
(iii) $x^2\frac{d^2y}{dx^2}(x) + 3x\frac{dy}{dx}(x) - 3y(x) = 0$,
(iv) $x^2\frac{d^2y}{dx^2}(x) - 4x\frac{dy}{dx}(x) + 6y(x) = x^6 - 3x^2$.

Exercise 3.9 (A boundary value problem) The goal of this exercise is to solve the following boundary value problem on the half-line numerically:

$$\frac{d^2f}{dr^2} + \frac{1}{r}\frac{df}{dr} + \left(1 - \frac{1}{r^2}\right)f = f^3, \quad f(0) = 0, \quad \lim_{r\to\infty} f(r) = 1. \tag{3.7}$$

The problem arises when one looks for spherically symmetric vortex solutions in the Ginzburg–Landau model of superfluids. It can be solved numerically using only standard numerical ODE solvers for initial value problems. However, in order to do this in practice we need to address two issues: the second order ODE (3.7) is singular at $r = 0$, and we only know the value of the function there. To start a numerical integration of a second order ODE we need to know the function and its derivative at a regular point. We proceed as follows:

(i) Find an approximate solution, valid near $r = 0$, of the form

$$f(r) = Ar^{\alpha}, \tag{3.8}$$

where you should determine α by inserting (3.8) into (1.2).
(ii) Pick a value for A and start the numerical integration at some small value r_0 (e.g., at $r_0 = 10^{-5}$), using the initial data

$$f(r_0) = Ar_0^{\alpha} \qquad f'(r_0) = \alpha A r_0^{\alpha-1}.$$

(iii) Vary A until the solution has the required limit for large r (in practice, $r = 10$ or $r = 15$ may be sufficient). This procedure for solving a boundary value problem is called a *shooting method.*

3.1.4 Reduction of order

Sometimes it is possible to reduce a second order differential equation to a first order differential equation. The most obvious case in which this is possible consists of equations of the form

$$\ddot{x} = f(t, \dot{x}).$$

Since the right hand side does not depend on x we can regard this as a first order differential equation for the function $v(t) = \dot{x}(t)$.

Reduction of order is less obvious for equations of the form

$$\ddot{x} = f(x, \dot{x}). \tag{3.9}$$

In that case, let $u = \dot{x}$ and regard u as a function of x. Then

$$\ddot{x} = \frac{d}{dt}\left(\frac{dx}{dt}\right) = \frac{du}{dt} = \frac{du}{dx}\frac{dx}{dt} = u\frac{du}{dx}.$$

Hence equation (3.9) becomes

$$u\frac{du}{dx} = f(x, u),$$

which is again of first order. For example, the equation

$$x\ddot{x} + (\dot{x})^2 = 0 \tag{3.10}$$

becomes

$$xu\frac{du}{dx} + u^2 = 0,$$

which is of first order and separable, with the general solution

$u(x) = A/x$, $A \in \mathbb{R}$. Recalling that $u = \dfrac{dx}{dt}$ and integrating again we obtain $\frac{1}{2}x^2 = At + B$ as an implicit formula for the general solution of (3.10).

The underlying reason for the reduction of order in both cases is related to symmetries of the equations in question. The systematic study of symmetry methods for reducing the order of an ODE is a very interesting subject in its own right, treated, for example, in Hydon [2] to which we refer the reader for further details.

Exercise 3.10 Solve the following initial value problems:
(i) $\dot{x}\ddot{x} = t, \quad x(1) = 2,\ \dot{x}(1) = 1.$
(ii) $\ddot{x} = \sin(x), \quad x(0) = \pi,\ \dot{x}(0) = 2.$ Does your method also work for the pendulum equation (1.1) with the same initial data?

Exercise 3.11 (i) Find the general solution of $\ddot{x} - \dot{x}t = 3$.
(ii) Show that the initial value problem $(\dot{x})^2\ddot{x} = 8x, \quad x(0) = \dot{x}(0) = 0$ has two distinct solutions. Why does the Picard–Lindelöf Theorem not apply?

Exercise 3.12 (Planetary motion and conservation laws) Let $\boldsymbol{x}(t)$ be the position vector of a planet relative to the Sun at time t. According to Newton's laws of motion and universal gravitation, the motion of the planet around the Sun is governed by the second order ODE (2.8) discussed in Example 2.1:

$$\ddot{\boldsymbol{x}} = -GM\frac{\boldsymbol{x}}{r^3}. \tag{3.11}$$

Here $r = |\boldsymbol{x}|$, M is the mass of the Sun and G Newton's gravitational constant. The problem of deducing the planet's trajectory from this differential equation is called the Kepler problem. This exercise guides you through a solution using conservation laws. You may need to look up the definitions of the following terms: scalar (or dot) product, vector (or cross) product, ellipse.
(i) Show that the following quantities are conserved (constant) during the motion of the planet:

$$E = \frac{1}{2}\dot{\boldsymbol{x}}\cdot\dot{\boldsymbol{x}} - \frac{GM}{r} \quad \text{(total energy)},$$

$$\boldsymbol{L} = \boldsymbol{x} \times \dot{\boldsymbol{x}} \quad \text{(angular momentum)},$$

$$\boldsymbol{K} = \dot{\boldsymbol{x}} \times \boldsymbol{L} - GM\frac{\boldsymbol{x}}{r} \quad \text{(Runge–Lenz vector)}.$$

(ii) Show further that $\boldsymbol{x}\cdot\boldsymbol{L} = \boldsymbol{K}\cdot\boldsymbol{L} = 0$ and deduce that the motion takes place in a plane orthogonal to $\boldsymbol{L}$, and that $\boldsymbol{K}$ is contained in that plane. Let ϕ be the angle between $\boldsymbol{x}$ and $\boldsymbol{K}$. Sketch the three vectors $\boldsymbol{L}, \boldsymbol{K}$ and $\boldsymbol{x}$, and mark ϕ in your diagram.
(iii) Finally, show that $\boldsymbol{x}\cdot\boldsymbol{K} = |\boldsymbol{L}|^2 - GMr$ and that the magnitude r of $\boldsymbol{x}$ depends on the angle ϕ according to

$$r = \frac{|\boldsymbol{L}|^2}{GM + |\boldsymbol{K}|\cos\phi}.$$

Deduce that the planet's orbit is an ellipse, with the Sun at one of the foci, provided $E < 0$. *Hint*: show $|\boldsymbol{K}|^2 = 2E|\boldsymbol{L}|^2 + (GM)^2$.

3.2 The oscillating spring

3.2.1 Deriving the equation of motion

Small and reversible deformations of elastic bodies are governed by a simple law which says, generally speaking, that the deformation is proportional to the force causing it. In the special case of an elastic spring, this rule is called Hooke's law. According to it, the force required to stretch a spring by an amount x from its equilibrium shape equals kx, where k is a constant which characterises the stiffness of the spring and which is called the *spring constant*; it has units N/m.† By Newton's third law, the spring exerts an equal and opposite force on the agent or object that is deforming it. This force, called the elastic restoring force, is thus

$$F = -kx. \tag{3.12}$$

Consider a spring hanging vertically as shown in Fig. 3.1. An object of mass m is attached to the free end of the spring, thus exerting a gravitational force of magnitude mg. According to Hooke's law (3.12), the spring will stretch by an amount l satisfying

$$mg = kl. \tag{3.13}$$

Suppose now that the spring is stretched by an additional amount x (with downward displacement counted positive) and that the object is moving. The forces acting on the object are

† See the end of the preface for our use of units.

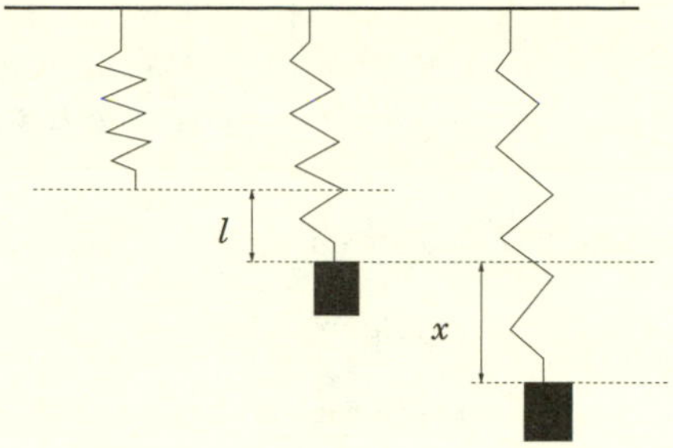

Fig. 3.1. The vertical spring

(i) the downward gravitational force of magnitude mg,
(ii) the elastic restoring force $-k(l+x)$,
(iii) air resistance,
(iv) any other force $f(t)$ exerted on the object.

The force due to air resistance is a damping force which is proportional to the velocity but acts in the opposite direction. It is given by $-r\frac{dx}{dt}$, where the *damping coefficient* r has units N s/m.

According to Newton's second law, the motion of the object is thus governed by the equation

$$m\frac{d^2x}{dt^2} = mg - k(l+x) - r\frac{dx}{dt} + f(t).$$

Re-arranging the terms and using (3.13) we arrive at the linear, inhomogeneous ODE

$$m\frac{d^2x}{dt^2} + r\frac{dx}{dt} + kx = f(t). \tag{3.14}$$

The dependence on the gravitational field strength g has dropped out of the equation - the spring's behaviour is the same on the Moon as it is on Earth! This is the sort of equation which we have learned to solve in the previous subsection. Here we interpret our solutions physically. The various parameters and functions are listed in Table 3.3 together with their physical interpretation.

In studying the equation (3.14), we start with the simplest situation and build up towards the general form. First, consider the case where the damping constant r is zero, and no additional force

Table 3.3. *Functions and parameters in the model for a mass on a spring. Downward displacement is counted as positive*

$x(t)$	displacement from equilibrium at time t
$\dot{x}(t)$	velocity at time t
$\ddot{x}(t)$	acceleration at time t
$f(t)$	external or driving force at time t
m	mass
r	damping coefficient
k	spring constant

f acts on the mass. With the abbreviation

$$\omega_0^2 = \frac{k}{m}, \tag{3.15}$$

the equation (3.14) thus becomes

$$\ddot{x} = -\omega_0^2 x. \tag{3.16}$$

This is the equation of *simple harmonic motion.* According to the discussion in Section 3.1.1, a fundamental set of solutions is given by $\{\cos(\omega_0 t), \sin(\omega_0 t)\}$, and the general solution is $x(t) = A\cos(\omega_0 t) + B\sin(\omega_0 t)$. Using basic trigonometry, we write it as

$$x(t) = R\cos(\omega_0 t - \phi), \tag{3.17}$$

where $R = \sqrt{A^2 + B^2}$ and $\tan\phi = B/A$. R is the furthest distance the object is displaced from the equilibrium position during the motion and is called the *amplitude* of the oscillation. Since cos is a periodic function with period 2π, the motion repeats itself after a time T which is such that $\omega_0 T = 2\pi$, i.e.,

$$T = \frac{2\pi}{\omega_0}.$$

This is called the *period* of the motion. The inverse $\nu = 1/T$ is the frequency and $\omega_0 = 2\pi\nu$ is an angular frequency called the *characteristic frequency* of the spring.

Exercise 3.13 An object of mass $m = 1$ kg is attached to a spring with spring constant $k = 16$ N/m. Give the equation of motion and determine the period of any solution. If the object is initially displaced downwards from the equilibrium by 1 m, and given an upward velocity of 4 m/s, find the downward displacement $x(t)$ at time $t > 0$. What is the amplitude of the motion? Sketch a graph of $x(t)$ and mark both the period and amplitude in your plot.

Exercise 3.14 Imagine a tunnel drilled through the centre of the Earth, from Spain to New Zealand. The total gravitational force exerted by the Earth's matter on a capsule travelling through the tunnel can be computed according to the following rule: *if the capsule is a distance r away from the centre of the Earth, the gravitational force it experiences is as if the Earth's matter inside the radius r was concentrated at the Earth's centre, and the matter outside the radius r was not there at all.*
(i) Find the mass M_E and radius R_E of the Earth by typing 'mass of Earth' and 'radius of Earth' into Google or otherwise.
(ii) Assuming the Earth to be a perfect ball with uniform mass density, compute the total mass of matter inside the ball of radius $r < R_E$ as a function of r.
(iii) Using the rule in italics and Newton's law of universal gravitation (3.11), compute the gravitational force experienced by the capsule when it is a distance r away from the centre of the Earth.
(iv) Derive the equation of motion governing the fall of the capsule to the centre of the Earth, neglecting air resistance and treating the capsule as a point particle; find its general solution.
(v) If the capsule is dropped into the tunnel in Spain with zero initial speed, find a formula for its position at time t seconds.
(vi) Using the Google calculator or otherwise, compute the time it takes the capsule to reach the centre of the Earth (in minutes), and the capsule's speed (in km/h) at that point. How long does it take until the capsule is in New Zealand, what is its speed when it gets there, and what is its average speed for the journey from Spain to New Zealand?

3.2.2 Unforced motion with damping

We continue to assume that there is no external force, but allow for a non-vanishing damping coefficient r. In terms of $\gamma = \frac{r}{m}$ and ω_0 defined in (3.15), the equation of motion (3.14) is thus

$$\frac{d^2x}{dt^2} + \gamma\frac{dx}{dt} + \omega_0^2 x = 0.$$

This is the general, constant coefficient, homogeneous linear ODE whose solutions we studied in Section 3.1.1. We now interpret those solutions physically, and sketch their graphs. The characteristic equation $\lambda^2 + \gamma\lambda + \omega_0^2 = 0$ has the roots

$$\lambda_1 = -\frac{\gamma}{2} + \sqrt{\frac{\gamma^2}{4} - \omega_0^2}, \quad \lambda_2 = -\frac{\gamma}{2} - \sqrt{\frac{\gamma^2}{4} - \omega_0^2}. \tag{3.18}$$

The discussion of the general solution and its physical interpretation is best organised according to the sign of $\gamma^2 - 4\omega_0^2$.

(i) *Overdamped case*: $\gamma^2 > 4\omega_0^2$

In this case, both roots λ_1 and λ_2 (3.18) are real and negative. According to Table 3.1, the general solution is

$$x(t) = Ae^{\lambda_1 t} + Be^{\lambda_2 t}. \tag{3.19}$$

Solutions are zero for at most one value of t (provided A and B do not both vanish) and tend to 0 for $t \to \infty$. A typical solution is shown in Fig. 3.2.

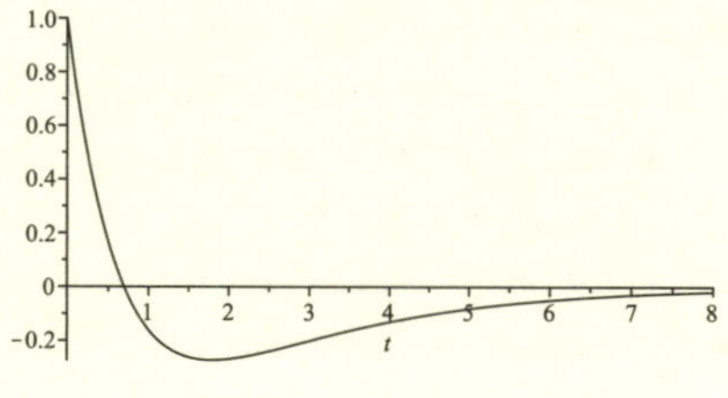

Fig. 3.2. The overdamped solution (3.19) for $\lambda_1 = -0.5, \lambda_2 = -1.5$, $A = -1, B = 2$

(ii) *Critically damped case*: $\gamma^2 = 4\omega_0^2$

In this case $\lambda_1 = \lambda_2 = -\gamma/2$ and the general solution is

$$x(t) = (A + Bt)e^{-\frac{\gamma}{2}t}, \tag{3.20}$$

which vanishes for at most one value of t (provided A and B do not both vanish) and tends to 0 for $t \to \infty$. A typical solution is shown in Fig. 3.3.

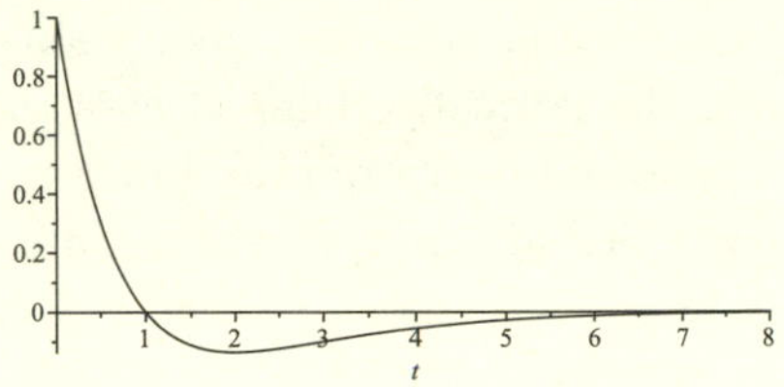

Fig. 3.3. Plot of the critically damped solution (3.20) for $\gamma = 2$, $A = 1, B = -1$

(iii) *Underdamped case:* $\gamma^2 < 4\omega_0^2$
We introduce the abbreviation

$$\beta = \sqrt{\omega_0^2 - \frac{\gamma^2}{4}} \tag{3.21}$$

so that $\lambda_1 = -\gamma/2 + i\beta$ and $\lambda_2 = -\gamma/2 - i\beta$. With the fundamental set of solutions from Table 3.1 we obtain the general solution

$$\begin{aligned} x(t) &= e^{-\frac{\gamma}{2}t}(A\cos(\beta t) + B\sin(\beta t)) \\ &= R\, e^{-\frac{\gamma}{2}t}\cos(\beta t - \phi), \end{aligned} \tag{3.22}$$

with R and ϕ defined as after equation (3.17). The solution oscillates and has infinitely many zeroes. The amplitude of the oscillations decreases exponentially with time. A typical solution is shown in Fig. 3.4.

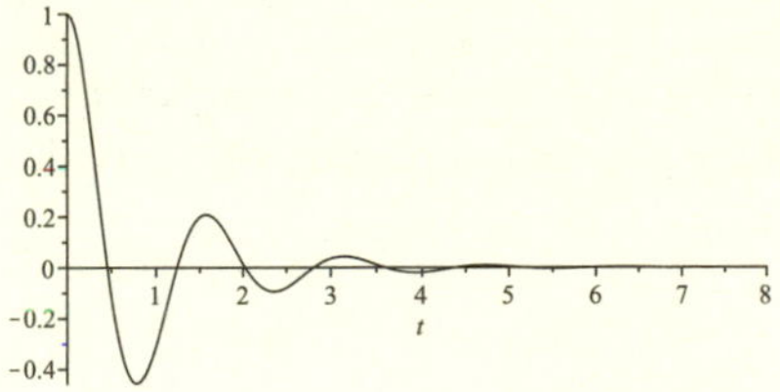

Fig. 3.4. Plot of the underdamped solution (3.22) for $\beta = 4, \gamma = 2$, $A = 1, B = 0.25$

The terminology 'underdamped', 'overdamped' and 'critically

damped' has its origin in engineering applications. The theory developed here applies, for example, to the springs that make up the suspension in cars. When perturbed from equilibrium, underdamped springs return to the equilibrium position quickly but overshoot. Overdamped springs take a long time to return to equilibrium. In the critically damped case the spring returns to the equilibrium position very quickly but avoids overshooting. In car suspensions, critically damped springs are most efficient at absorbing bumps.

Exercise 3.15 An object of mass 1 kg is attached to a spring with spring constant 1 N/m and is immersed in a viscous fluid with damping constant 2 Nsec/m. At time $t = 0$ the object is lowered $\frac{1}{4}$ m and given an initial velocity of 1 m/sec in the upward direction. Find the subsequent motion of the object and show that it will overshoot its equilibrium position once before returning to equilibrium. Sketch the position of the object as a function of time.

3.2.3 Forced motion with damping

We turn to the general case, where all terms in the equation of motion (3.14) play a role, and focus on the situation where f is a periodic function. This case is important in applications and also provides a key to understanding a more general situation via the tool of the Fourier transform. Suppose therefore that

$$f(t) = f_0 \cos(\omega t), \tag{3.23}$$

where f_0 and ω are constants which characterise, respectively, the amplitude and angular frequency of the driving force. Then the equation of motion (3.14) is equivalent to

$$\frac{d^2x}{dt^2} + \gamma\frac{dx}{dt} + \omega_0^2 x = \frac{f_0}{m}\cos(\omega t), \tag{3.24}$$

where we again use $\gamma = \dfrac{r}{m}$ and ω_0 given in (3.15), and assume that $r \neq 0$. The quickest way to solve this is to use complex numbers and Table 3.2. Since $\cos\omega t$ is the real part of $\exp(i\omega t)$

we first solve

$$\frac{d^2x}{dt^2} + \gamma\frac{dx}{dt} + \omega_0^2 x = \frac{f_0}{m}e^{i\omega t}, \tag{3.25}$$

and then take the real part of the solution. Since $\gamma \neq 0$, $i\omega$ is not a solution of the characteristic equation of the homogeneous equation. Therefore we look for a complex solution $x(t) = C\exp(i\omega t)$ of the inhomogeneous equation (3.25). Inserting gives

$$Ce^{i\omega t}(-\omega^2 + i\gamma\omega + \omega_0^2) = \frac{f_0}{m}e^{i\omega t}.$$

Dividing by $\exp(i\omega t)$ and solving for C, we deduce

$$C = \frac{(f_0/m)}{-\omega^2 + i\gamma\omega + \omega_0^2} = Re^{-i\phi},$$

where we have parametrised C in terms of the modulus R and the phase ϕ (which is the negative of the conventional argument in the modulus-argument form of a complex number). We are particularly interested in their functional dependence on ω and therefore write

$$R(\omega) = \frac{(f_0/m)}{\sqrt{(\omega_0^2 - \omega^2)^2 + \gamma^2\omega^2}}, \qquad \tan\phi(\omega) = \frac{\gamma\omega}{\omega_0^2 - \omega^2}. \tag{3.26}$$

Taking the real part of $x(t) = C\exp(i\omega t)$, we obtain the solution

$$x_p(t) = R(\omega)\cos(\omega t - \phi(\omega)). \tag{3.27}$$

The function $R(\omega)$ gives the ω-dependent amplitude of the forced motion. Its typical shape is shown in Fig. 3.5.

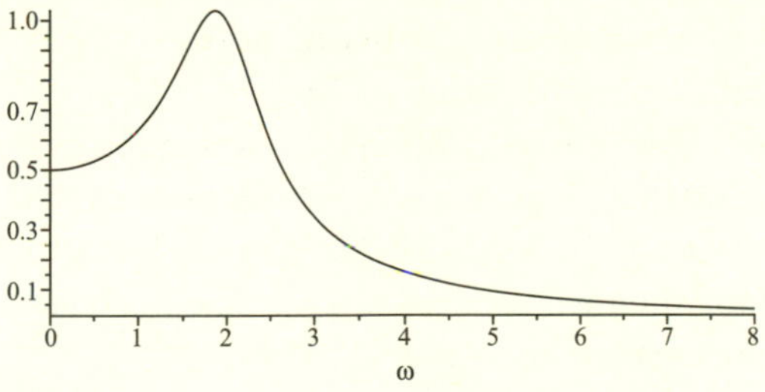

Fig. 3.5. The amplitude (3.26) for $f_0/m = 2, \omega_0 = 2, \gamma = 1$ (units suppressed)

Looking at (3.26), we note that the particular solution has special properties when $\omega = \omega_0$: the amplitude R is large, and $\tan\phi(\omega_0) = \infty$, so $\phi(\omega_0) = \pi/2$ and the phase of the particular solution (3.27) lags behind that of the driving force by $\pi/2$.

Definition 3.1 The coincidence of the driving frequency and the characteristic frequency is called *resonance*. In symbols, the resonance condition is $\omega = \omega_0$.

The amplitude R is not actually maximal at resonance. However, the rate at which energy is transferred to the oscillating mass *is* maximal when $\omega = \omega_0$. This fact is explained and explored in Exercise 3.18, and gives the physical motivation for the definition (3.1). It also contains an important message for applications of driven oscillations in engineering - or everyday life. If you want to transfer energy to an oscillating system, it is best to do so with the characteristic frequency of that system. This is a rule which we intuitively obey when pushing another person on a swing, for example: pushing with the natural frequency of the swing leads to a large amplitude of oscillations, whereas pushing with a very different frequency would only produce small wobbles.

The example of the swing also shows that the energy transferred at resonance can be destructive. If we keep pushing at the resonance frequency we may send the person on the swing into a circular orbit! The danger of destructive energy transfer at resonance is also the reason why soldiers break their step when marching across a bridge: if their step frequency is equal to one of the natural oscillation frequencies of the bridge, the periodic force they exert on the bridge when marching in step might destroy it. We return to resonance problems in bridges briefly in the next subsection and, in more detail, in Project 5.3.

The general solution of (3.24) is a linear combination of the particular solution (3.27) and the general solution of the homogeneous equations discussed in Section 3.2.2. Let us, for definiteness, consider the underdamped case $\gamma^2 < 4\omega_0^2$. Then the general solution of the inhomogeneous equation (3.24) is

$$x(t) = e^{-\frac{\gamma}{2}t}(A\cos(\beta t) + B\sin(\beta t)) + x_p(t),$$

with β defined in (3.21) and x_p in (3.27). After a while, the solution is dominated by the particular solution (3.27), called *steady state solution* in physics and engineering; it has the same frequency as the driving term (3.23). The part of the solution which solves the homogeneous equation is called the *transient solution*; it tends to zero for $t \to \infty$.

3.2.4 Forced motion without damping

In the absence of damping ($r = 0$ and hence $\gamma = 0$) the equation (3.24) simplifies to

$$\frac{d^2x}{dt^2} + \omega_0^2 x = \frac{f_0}{m}\cos(\omega t).$$

First consider the case $\omega \neq \omega_0$, i.e., $i\omega$ is not a root of the characteristic equation. Using Table 3.2 we find the particular solution $x_p(t) = \frac{(f_0/m)}{(\omega_0^2-\omega^2)}\cos(\omega t)$ and therefore the general solution

$$x(t) = A\cos(\omega_0 t) + B\sin(\omega_0 t) + \frac{f_0/m}{(\omega_0^2 - \omega^2)}\cos(\omega t), \qquad (3.28)$$

which is a fairly complicated superposition of oscillations of different frequencies. It remains bounded as $t \to \infty$.

Next consider the case of resonance. The driving frequency equals the spring's characteristic frequency, i.e., $\omega = \omega_0$. Again we think of $\cos(\omega t)$ as the real part of $\exp(i\omega t)$ and study

$$\frac{d^2x}{dt^2} + \omega^2 x = \frac{f_0}{m}e^{i\omega t}.$$

Since $i\omega$ is a solution of the characteristic equation, we try $x(t) = Ct\exp(i\omega t)$ in accordance with Table 3.2. We find

$$2i\omega Ce^{i\omega t} = \frac{f_0}{m}e^{i\omega t} \Rightarrow C = \frac{f_0}{2mi\omega}.$$

Taking the real part of

$$x(t) = Ct\exp(i\omega t) = \frac{f_0 t}{2m\omega}(-i\cos(\omega t) + \sin(\omega t))$$

we obtain a particular solution $x_p(t) = \frac{f_0 t}{2m\omega}\sin(\omega t)$, and thus the general solution

$$x(t) = A\cos(\omega t) + B\sin(\omega t) + \frac{f_0 t}{2m\omega}\sin(\omega t). \qquad (3.29)$$

The remarkable property of the particular solution is that its amplitude grows linearly with time and becomes infinite as $t \to \infty$. In real life, ever increasing oscillations mean that the oscillating system (be it a spring or a more complicated object – such as a building) will break. This can lead to very dramatic manifestations of resonance, as for example in the collapse of the Tacoma Narrows bridge in the USA in 1940. Video footage of this collapse can be found on the internet (e.g., on youtube).

The general solution (3.28) of the driven, undamped oscillations away from resonance looks quite different from the general solution (3.29) at resonance. One can nonetheless show that, for fixed initial conditions, the solution depends smoothly on the parameters ω and ω_0. You are asked to do this in Exercise 3.20, as an example of studying continuous dependence of solutions on parameters. According to our discussion in Section 1.1.2, this is a required feature of a well-posed problem.

Exercise 3.16 An object of mass 1 kg is attached to a spring with spring constant 16 Nm^{-1} and allowed to oscillate freely.
(i) Neglecting air resistance, find the angular frequency and the period of the oscillations.
(ii) The object is acted upon by the external force $f_1(t) = f_0 \cos 2t$. Give the equation of motion and find its general solution.
(iii) The external force is changed to $f_2(t) = f_0 \cos 4t$. Find a particular solution of the new equation of motion.
(iv) Now both forces f_1 and f_2 act on the object. Find a particular solution of the resulting equation of motion.

Exercise 3.17 An object of mass 4 kg is attached to a spring with spring constant 64 N/m and is acted on by an external force $f(t) = A\cos^3(pt)$ in the downward direction. Ignoring air resistance, find all values of p at which resonance occurs.

Exercise 3.18 The energy of an object of mass m attached to a spring with the spring constant k is $E = \frac{1}{2}m\dot{x}^2 + \frac{1}{2}kx^2$, with x and $\dot{x}$ explained in Table 3.3. Suppose the object is immersed in a viscous substance with damping coefficient r and acted on by an external force $f(t) = f_0 \cos \omega t$.
(i) Use the equation of motion (3.24) to deduce that the energy

changes with time according to $\frac{dE}{dt} = -r\dot{x}^2 + \dot{x}f$. The term $-r\dot{x}^2$ is called the dissipated power, and the term $P = \dot{x}f$ is called the power input.
(ii) Considering only the particular solution (3.27), compute the average power input per cycle

$$\bar{P}(\omega) = \frac{1}{T}\int_0^T P(t)dt,$$

where $T = \frac{2\pi}{\omega}$. Show that this function has an absolute maximum at the resonance frequency $\omega_0 = \sqrt{k/m}$.

Exercise 3.19 An object of mass 1 kg is attached to a spring with the spring constant $k = 4$ Nm^{-1}. The spring is immersed in a viscous liquid with damping constant $r = 1$ N sec m^{-1} and an external force $f(t) = 2\cos\omega t$ N is applied to the mass, where $\omega \geq 0$.
(i) Write down the differential equation governing the motion of the object and find the steady state solution. Give its amplitude R as a function of ω. Find the frequency $\omega_{\max}$ at which the amplitude $R(\omega)$ attains its maximum value. Sketch the function $R(\omega)$ and mark the value of $\omega_{\max}$ in your diagram. Also mark the resonance frequency in your diagram.
(ii) Consider the same spring and the same external force, but now with weaker damping $r_1 = 0.5$ N sec m^{-1} and very weak damping $r_2 = 0.1$ N sec m^{-1}. For both values of the damping coefficient, find, as a function of ω, the amplitude $R(\omega)$ of the steady state solution and the phase $\phi(\omega)$ of the steady state solution relative to the external force f. Plot (either by hand or using a computer) R and ϕ as a function of ω. Also plot the steady state solutions *at resonance* for both $r_1 = 0.5$ N sec m^{-1} and $r_2 = 0.1$ N sec m^{-1} as functions of time. Plot the driving force as a function of time in the same diagram.

Exercise 3.20 Determine the constants A and B in both (3.28) and (3.29) so that both solutions satisfy the initial conditions $x(0) = \dot{x}(0) = 0$. Denote the resulting solution of (3.28) by $x_{\omega_0}(t)$ and the resulting solution of (3.29) by $x_{\text{res}}(t)$. Show that, for fixed t, $\lim_{\omega_0 \to \omega} x_{\omega_0}(t) = x_{\text{res}}(t)$.

4
Geometric methods

4.1 Phase diagrams

4.1.1 Motivation

In this chapter we study ways of gaining qualitative insights into solutions of a given ODE (or system of ODEs) without actually solving the ODE (or the system). We already studied one method for doing this in Section 1.4, where we sketched direction fields for ODEs and were able to deduce qualitative properties of solutions, like their asymptotic behaviour. If you look again at the equation

$$\frac{dx}{dt} = x(x-1)(x-2), \tag{4.1}$$

studied in Example 1.2, you may notice that the direction field for this example, shown in Fig. 1.7, contains much redundant information: since the right hand side of the ODE (4.1) does not depend on t, there is little point in sketching the arrows in the (t, x)-plane. Instead, we could drop the t-coordinate and simply draw arrows on the x-axis to show if the solution through a given point x is moving up or down. After turning the x-axis on its side, we then obtain the picture shown in Fig. 4.1 (where 'up' has become 'right' and 'down' has become 'left'). In the picture we have also marked the constant solutions $x = 0$, $x = 1$ and $x = 2$ with fat dots. These, together with the arrows, give us a qualitative but accurate idea of how solutions for given initial conditions behave. If the initial value coincides with a fat dot, the corresponding solution is constant. In a region where the arrows point right, the solution grows; in a region where the arrows point left, it decreases. Solutions can never reach or pass through fat dots because that would violate the local uniqueness of solutions guaranteed by the Picard–Lindelöf Theorem (whose assumptions are satisfied in the case (4.1)).

The picture shown in Fig. 4.1 is an example of a *phase diagram.* It is a useful tool for obtaining qualitative insights into solutions of ODEs which do not explicitly depend on the independent variable t. In the next section we define, study and use phase diagrams.

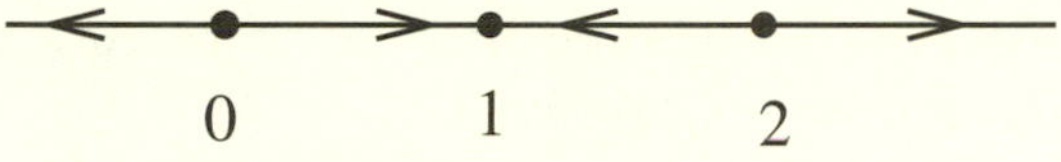

Fig. 4.1. Phase diagram for the differential equation (4.1)

4.1.2 Definitions and examples

A system of ordinary differential equations which can be written in the form

$$\dot{\boldsymbol{x}} = \boldsymbol{f}(\boldsymbol{x}), \tag{4.2}$$

with $\boldsymbol{x} : \mathbb{R} \to \mathbb{R}^n$ and $\boldsymbol{f} : \mathbb{R}^n \to \mathbb{R}^n$, is called *autonomous.* The space where solutions $\boldsymbol{x}$ take their values is called the *phase space* of the system (4.2). Here we consider the case where the phase space is all of $\mathbb{R}^n$, and focus on the cases $\mathbb{R}$ and $\mathbb{R}^2$. In general, the phase space could be a subset of $\mathbb{R}^n$ or, in fact, a differentiable manifold.

The vector $\boldsymbol{f}(\boldsymbol{x})$ is equal to the velocity $\dot{\boldsymbol{x}}$ of the solution curve through $\boldsymbol{x}$. The vector field $\boldsymbol{f}$ is therefore called the *phase velocity vector field*, or sometimes simply the vector field associated with (4.2). The *trajectory* of a solution $\boldsymbol{x}$ of (4.2) is the set of all points reached by $\boldsymbol{x}(t)$ for some value of t. We define the *phase diagram* of (4.2) to be the phase space $\mathbb{R}^n$ with trajectories of (4.2) drawn through each point. The phase diagram shows all possible trajectories of an autonomous differential equation. In practice, we only sketch a few trajectories. Alternatively, we sketch the vector field, i.e., we draw arrows representing $\boldsymbol{f}(\boldsymbol{x})$ at points $\boldsymbol{x}$ in the phase space.

Points where the vector field $\boldsymbol{f}$ vanishes play an important role in understanding the qualitative behaviour of solutions and are called *equilibrium points.* Every equilibrium point $\boldsymbol{x}^*$ is itself the trajectory of a constant solution $\boldsymbol{x}(t) = \boldsymbol{x}^*$ for all $t \in \mathbb{R}$ since

$$\frac{d\boldsymbol{x}}{dt} = \boldsymbol{f}(\boldsymbol{x}^*) = 0.$$

An equilibrium point is called *stable* if, when you start close enough to it, you stay close to it:

Definition 4.1 An equilibrium point $\boldsymbol{x}^*$ is called stable if for every $\epsilon > 0$ there is a $\delta > 0$ so that

$$|\boldsymbol{x}_0 - \boldsymbol{x}^*| < \delta \Rightarrow |\boldsymbol{x}(t) - \boldsymbol{x}^*| < \epsilon, \quad t \geq 0 \tag{4.3}$$

for every solution $\boldsymbol{x}$ of (4.2) with $\boldsymbol{x}(0) = \boldsymbol{x}_0$.

An equilibrium point is called *unstable* if it is not stable. An equilibrium point is called *attracting* if, when you start close enough, you actually tend to it. The following definition makes this notion precise.

Definition 4.2 An equilibrium point $\boldsymbol{x}^*$ is called attracting if there is a $\delta > 0$ so that

$$|\boldsymbol{x}_0 - \boldsymbol{x}^*| < \delta \Rightarrow \boldsymbol{x}(t) \to \boldsymbol{x}^* \quad \text{as} \quad t \to \infty \tag{4.4}$$

for every solution $\boldsymbol{x}$ of (4.2) with $\boldsymbol{x}(0) = \boldsymbol{x}_0$.

We study examples of stable points which are not attracting below. In one-dimensional systems, attracting points are automatically stable. Perhaps surprisingly, there exist attracting points which are not stable in two or more dimensions. These issues are investigated in detail in Exercises 4.3 and 4.4. For now, we introduce a term for equilibrium points which are both stable and attracting:

Definition 4.3 An equilibrium point $\boldsymbol{x}^*$ is called asymptotically stable if it is stable and attracting.

Example 4.1 Sketch phase diagrams for $\dot{x} = f(x)$ with

$$\text{(i)}\, f(x) = x, \quad \text{(ii)}\, f(x) = 1 - x^2, \quad \text{(iii)}\, f(x) = -x^2. \tag{4.5}$$

In each case, determine all equilibrium points and their nature.

In all three cases, the phase space is the entire real line. We first identify the equilibrium points and mark them with a black dot. The equilibrium points divide the line into separate regions. In each region we pick points x and then draw arrows in the direction of $f(x)$: if $f(x)$ is positive the arrow points to the right, if it is negative the arrow points to the left. The resulting phase diagrams are shown in Fig. 4.2. The arrows allow us to determine the stability of the equilibrium points. The diagram for (i) shows an unstable equilibrium at $x = 0$, the diagram for (ii) has an unstable equilibrium point at $x = -1$ and a stable equilibrium point at $x = 1$. The diagram for (iii) shows an unstable equilibrium point at $x = 0$. This point is 'less unstable' than the equilibrium point in (i) in the sense that the points to the right of $x^* = 0$ will stay close to $x^* = 0$. It is therefore sometimes called *semistable*. □

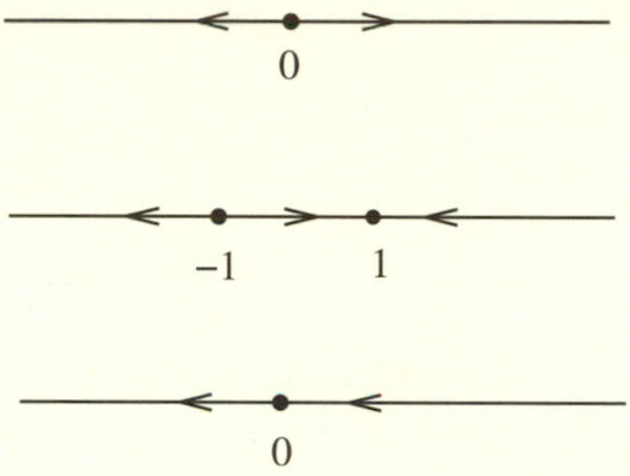

Fig. 4.2. The phase diagrams for the one-dimensional systems (4.5): (i) at the top (ii) in the middle, (iii) at the bottom

Exercise 4.1 Draw phase diagrams for the equations

$$\text{(i) } \dot{x} = x - x^2, \qquad \text{(ii) } \dot{x} = \sin(x).$$

Exercise 4.2 Show that the one-dimensional system

$$\dot{x} = \begin{cases} -x^2 & \text{if} \quad x < 0 \\ 0 & \text{if} \quad 0 \le x \le 1 \\ -(1-x)^2 & \text{if} \quad x > 1 \end{cases}$$

has many stable points which are not asymptotically stable.

Exercise 4.3 Consider the equation $\dot{x} = f(x)$, where $f : \mathbb{R} \to \mathbb{R}$ is continuously differentiable.

(i) Show that, for any solution $x(t)$, if $f(x(t_0)) = 0$ for some t_0, then $f(x(t)) = 0$ for all $t \in \mathbb{R}$.

(ii) Show also that, for any solution $x(t)$, if $f(x(t_0)) > 0$ for some t_0, then $f(x(t)) > 0$ for all $t \in \mathbb{R}$; show that the same holds when $>$ is replaced with $<$.

(iii) Use (i) and (ii) to show that if an equilibrium point x^* is attracting then it must also be stable.

Next we consider some two-dimensional examples.

Example 4.2 Sketch the phase diagram for the equation of simple harmonic motion $\ddot{x} = -x$.

The equation may be written as the system

$$\frac{dx}{dt} = y, \quad \frac{dy}{dt} = -x.$$

The only equilibrium point is $(x, y) = (0, 0)$ and it is easy to check that

$$\frac{d}{dt}\left(x^2 + y^2\right) = 0,$$

so that trajectories are determined by $x^2 + y^2 = c$ (we give a systematic way of finding the equation for trajectories in Example 4.3 below). Hence the trajectories are circles centred on $(0, 0)$ and the phase diagram is as shown on the left in Fig. 4.3. Since $\frac{dx}{dt} > 0$ when $y > 0$ we deduce that the flow on the phase trajectories in the upper half-plane is to the right, which gives the direction of the arrows in the picture. We can also sketch the solution $x(t)$ directly from the phase diagram. To do this, we follow the change in x as we flow once round a circle in the phase diagram. The result is shown in the middle and right of Fig. 4.3, with corresponding points on the phase trajectory and the plot of $x(t)$ indicated by the numbers 1–4. □

Example 4.3 Sketch the phase diagram of $\ddot{x} = -x^3$.

The equation may be written as the system

$$\frac{dx}{dt} = y, \quad \frac{dy}{dt} = -x^3. \tag{4.6}$$

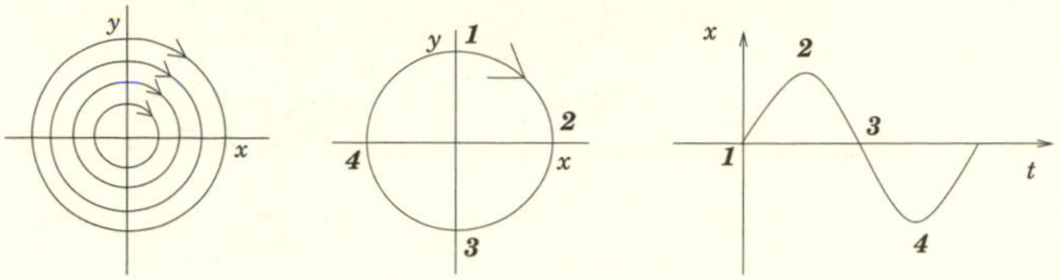

Fig. 4.3. The phase diagram for simple harmonic motion: Phase trajectories and the correspondence between phase trajectory and displacement $x(t)$

Again the only equilibrium point is $(x, y) = (0, 0)$. In order to find the trajectory we can eliminate t by considering the quotient

$$\frac{dy}{dx} = \frac{\frac{dy}{dt}}{\frac{dx}{dt}} = -\frac{x^3}{y}, \quad \text{provided } y \neq 0.$$

Separating variables and integrating, we find $y^2 = -\frac{1}{2}x^4 + c$. Alternatively, we can avoid the division by y if we start with the trivial observation that $\dot{x}\dot{y} - \dot{y}\dot{x} = 0$ and replace $\dot{x}$ in the first term by y and $\dot{y}$ in the second term by $-x^3$ to conclude that

$$y\frac{dy}{dt} + x^3\frac{dx}{dt} = \frac{d}{dt}\left(\frac{1}{2}y^2 + \frac{1}{4}x^4\right) = 0 \Leftrightarrow y^2 + \frac{1}{2}x^4 = c.$$

For the trajectory through the point $(0, a > 0)$ we have

$$y^2 = -\frac{1}{2}x^4 + a^2. \tag{4.7}$$

As we increase x from 0, y^2 and hence also y decrease until y reaches 0 at the point when $x = b = (2a^2)^{1/4}$. The equation for the trajectory (4.7) is reflection symmetric with respect to the x- and y-axis, i.e., if (x, y) is a point on the trajectory then so are $(-x, y)$, $(x, -y)$ and $(-x, -y)$. Thus we obtain the full trajectory shown on the left of Fig. 4.4. Determining the direction of the flow as in Example 4.2 we arrive at the phase diagram shown on the right of Fig. 4.4. Note that all solutions are periodic. □

Example 4.4 (The pendulum equation) The mathematical pendulum is an idealised pendulum. It consists of a rigid rod of length ℓ, assumed massless, which is pivoted at one end and has a bob of mass m attached at the other. The bob is subjected to the force of gravity near the Earth's surface. The pendulum and

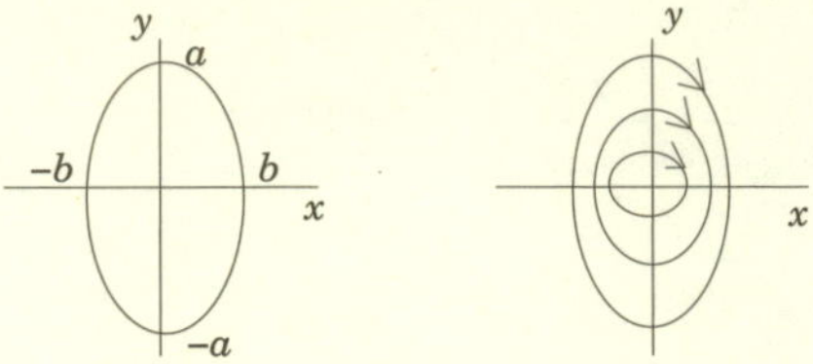

Fig. 4.4. Phase trajectory and phase diagram for $\ddot{x} = -x^3$

the forces acting on the bob are sketched in Fig. 4.5, which also shows the resolution of the downward gravitational force into a component in the direction of the rod and a component perpendicular to it. Considering the motion perpendicular to the rod, we deduce the pendulum equation:

$$ml\ddot{x} = -mg\sin(x) \quad \Leftrightarrow \ddot{x} = -\frac{g}{\ell}\sin(x). \tag{4.8}$$

Sketch the phase diagram of the pendulum equation for the case $g/\ell = 1$.

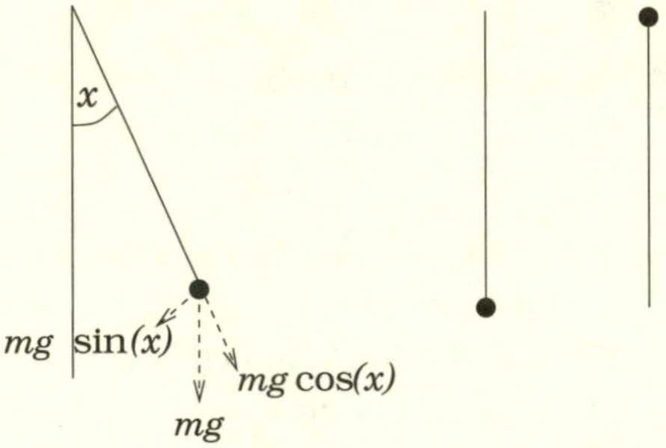

Fig. 4.5. Forces and equilibrium positions for the pendulum

Remark If x is small, we can approximate $\sin x \approx x$. Then the equation (4.8) simplifies to $\ddot{x} = -\frac{g}{\ell}x$, which is the simple harmonic oscillator equation.

The pendulum equation may be written as the system

$$\frac{dx}{dt} = y, \quad \frac{dy}{dt} = -\sin(x). \tag{4.9}$$

The equilibrium points are $(x, y) = (n\pi, 0)$ for $n \in \mathbb{Z}$, corresponding to equilibrium positions 'down' and 'up', also shown in Fig. 4.5. Any trajectory must satisfy

$$y\frac{dy}{dt} = -\sin(x)\,\frac{dx}{dt} \;\Leftrightarrow\; y^2 = 2\cos(x) + c.$$

Thus the trajectory passing through $(0, a)$, where $a > 0$, has the equation

$$y^2 = 2(\cos(x) - 1) + a^2. \tag{4.10}$$

The function $2(\cos(x) - 1)$ decreases from 0 to -4 as x increases from 0 to π. The form of the trajectory thus depends upon whether $a^2 < 4$, $a^2 > 4$ or $a^2 = 4$, and we need to consider theses cases separately.

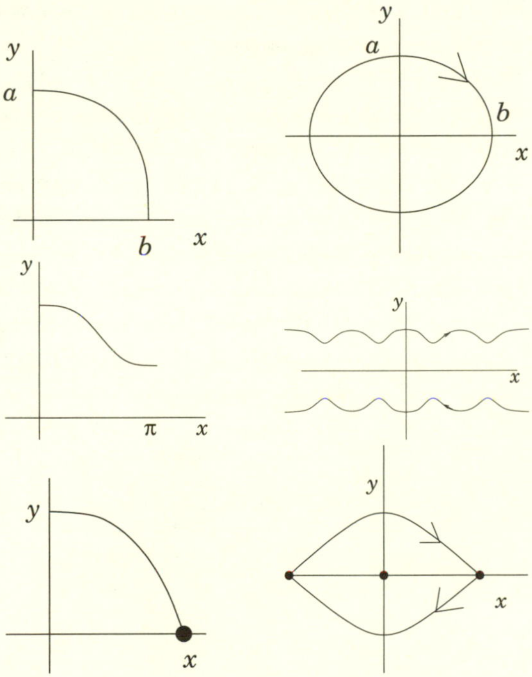

Fig. 4.6. Sketching the phase diagram for the pendulum equation (4.9). Explanation in the main text

(i) $a^2 < 4$: As x increases from 0 towards π, y^2 decreases until

it reaches $y^2 = 0$ at a value of x such that $\cos x = 1 - a^2/2$; we denote this value by b. The corresponding trajectory in the first quadrant is shown on the top left in Fig. 4.6. Since equation (4.10) for the trajectory is symmetric with respect to the x- and y-axis, it follows that if (x, y) is on a trajectory, so are the points $(-x, y), (x, -y), (-x, -y)$ obtained by reflections in the x- and y-axis. Thus the full trajectory is as shown on the top right of Fig. 4.6. The corresponding solution is periodic: the pendulum oscillates between $x = \pm b$ (like the simple harmonic oscillator).
(ii) $a^2 > 4$: As x increases from 0 to π, y^2 decreases from a^2 to the minimum value $a^2 - 4$, attained when $x = \pi$. Hence we obtain the trajectory shown on the middle left of Fig. 4.6. The gradient at $x = \pi$ is $\frac{dy}{dx} = -\sin(\pi)/\sqrt{a^2 - 4} = 0$. By reflection symmetry in x and y and 2π-periodicity in x we obtain the trajectories shown in the middle right of Fig. 4.6. These trajectories correspond to solutions where the pendulum carries out complete revolutions.
(iii) $a^2 = 4$: Now (4.10) becomes $y^2 = 2(\cos x + 1)$. Hence, as $x \to \pi$, $y \to 0$, so that the trajectory approaches the equilibrium point $(\pi, 0)$ as shown in the bottom left of Fig. 4.6. In terms of the picture on the right of Fig. 4.5, this trajectory corresponds to the pendulum starting near the 'down' equilibrium position and approaching the 'up' equilibrium position as $t \to \infty$. By symmetry, we obtain the trajectories shown in the bottom right of Fig. 4.6. Since all the equations for trajectories are periodic in x with period 2π, we finally obtain the full phase diagram of the mathematical pendulum shown in Fig. 4.7.

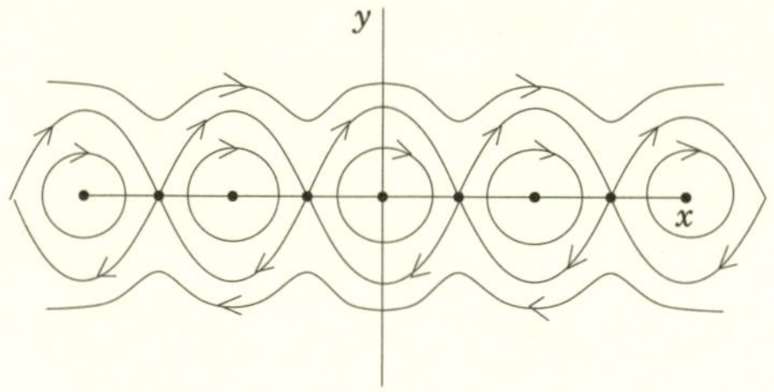

Fig. 4.7. The phase diagram for the pendulum equation (4.9)

Exercise 4.4 Consider polar coordinates (r, ϕ) in the plane. They have the ranges $0 \leq r < \infty$ and $0 \leq \phi < 2\pi$, and are related to Cartesian coordinates according to $x = r\cos\phi$, $y = r\sin\phi$ (compare also Exercise 2.5). Sketch the phase diagram for the system

$$\dot{r} = r(1-r), \quad \dot{\phi} = r(\cos\phi - 1)$$

and show that the equilibrium point $(x, y) = (1, 0)$ is attracting but not stable.

4.1.3 Phase diagrams for linear systems

In this section, we study in detail phase diagrams for linear homogeneous systems

$$\dot{\boldsymbol{x}} = A\boldsymbol{x}. \tag{4.11}$$

For clarity and simplicity we focus on the systems of size two, i.e., on the case where A is a constant 2×2 matrix and $\boldsymbol{x} : \mathbb{R} \mapsto \mathbb{R}^2$. We use the eigenvalue method of Section 2.4.2 to find solutions and then sketch the solutions in the phase diagram. In doing so, we need to distinguish cases and this organises our discussion in terms of the nature of the eigenvalues of the matrix A. We will find that, for each type of 2×2 matrix, there is an associated type of phase diagram.

Since phase diagrams are intended to help us understand differential equations in the absence of exact solutions it may seem odd to consider the solvable equation (4.11), solve it and then sketch the solutions. However, the investigation of linear systems is useful in the study of non-linear autonomous systems of the form

$$\dot{\boldsymbol{x}} = \boldsymbol{f}(\boldsymbol{x}), \tag{4.12}$$

where $\boldsymbol{f} : \mathbb{R}^2 \to \mathbb{R}^2$ is a non-linear function. Such systems cannot, in general, be solved exactly. However, one can learn much about the qualitative behaviour of solutions by linearising about equilibrium points. We will study the technique of linearisation in Section 4.2, where we also discuss its validity and accuracy.

Returning to the linear system (4.11) we introduce the notation

$$\boldsymbol{x} = \begin{pmatrix} x \\ y \end{pmatrix} \text{ and } A = \begin{pmatrix} a & b \\ c & d \end{pmatrix}, \quad a, b, c, d \in \mathbb{R}, \tag{4.13}$$

so that (4.11) is explicitly given by

$$\begin{aligned} \dot{x} &= ax + by \\ \dot{y} &= cx + dy. \end{aligned} \tag{4.14}$$

Clearly, $(0, 0)$ is always an equilibrium point of (4.11). If both eigenvalues of A are non-zero (so that A is invertible) then $(0, 0)$ is the only equilibrium point, but if one eigenvalue vanishes we have a whole line of equilibrium points.

We now discuss the nature of these equilibrium systematically. In our discussion we make use of the results of Section 2.4.2, where we learned how to construct the general solution of (4.11) or, equivalently, of (4.14) by making the ansatz $\boldsymbol{x}(t) = e^{\lambda t}\boldsymbol{v}$, with λ an eigenvalue of A and $\boldsymbol{v}$ the corresponding eigenvector $\boldsymbol{v}$.

(a) **Distinct, real and negative eigenvalues**: $\lambda_1 < \lambda_2 < 0$. In this case the general solution of (4.14) is

$$\boldsymbol{x}(t) = c_1 e^{\lambda_1 t}\boldsymbol{v}_1 + c_2 e^{\lambda_2 t}\boldsymbol{v}_2, \tag{4.15}$$

where $\boldsymbol{v}_1$ and $\boldsymbol{v}_2$ are the eigenvectors with eigenvalues λ_1 and λ_2. Suppose we have $c_1 > 0, c_2 = 0$. The trajectory corresponding to this solution is along the ray in the direction of $\boldsymbol{v}_1$, with the arrow pointing towards the origin. If $c_1 < 0, c_2 = 0$, the trajectory is along the ray in the direction of $-\boldsymbol{v}_1$, with the arrow pointing towards the origin. Similarly, if $c_1 = 0, c_2 > 0\,(< 0)$, the trajectory is along the ray in the direction of $\boldsymbol{v}_2$, $(-\boldsymbol{v}_2)$, with the arrow pointing towards the origin.

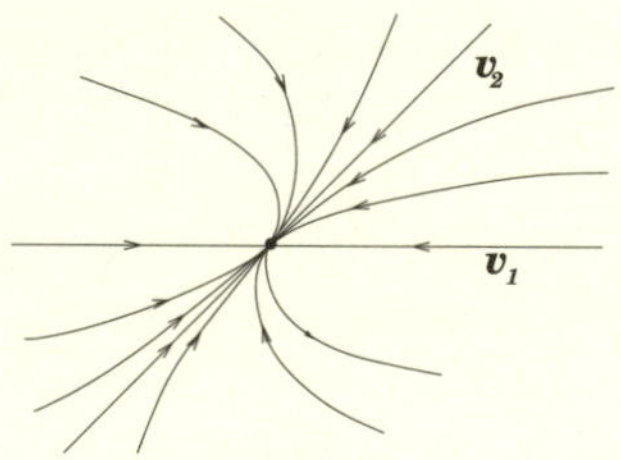

Fig. 4.8. Typical phase diagram for a stable node

Now consider the trajectory corresponding to the general solution (4.15). Clearly the corresponding trajectory heads into the equilibrium point $(0, 0)$ as $t \to \infty$. For large t, the term $c_2 e^{\lambda_2 t}\boldsymbol{v}_2$

dominates and so the direction of approach to $(0,0)$ is almost parallel to the direction of $\boldsymbol{v}_2$. Similarly, when $t \to -\infty$, the first term $c_1 e^{\lambda_1 t}\boldsymbol{v}_1$ dominates, and the direction of the trajectory approaches that of $\boldsymbol{v}_1$. The resulting phase diagram is shown in Fig. 4.8. The origin $(0,0)$ is called a *stable node.*

(b) **Distinct, real and positive eigenvalues**: $0 < \lambda_1 < \lambda_2$. The general solution is again given by (4.15). All (non-constant) trajectories move out from the equilibrium point $(0,0)$. At $t \to \infty$, the term $c_2 e^{\lambda_2 t}\boldsymbol{v}_2$ dominates and so the direction of any trajectory approaches that of $\boldsymbol{v}_2$. As $t \to -\infty$, the term $c_1 e^{\lambda_2 t}\boldsymbol{v}_1$ dominates and so the direction of any trajectory approaches that of $\boldsymbol{v}_1$. The phase diagram has the form shown in Fig. 4.9. The point $(0,0)$ is called an *unstable node.*

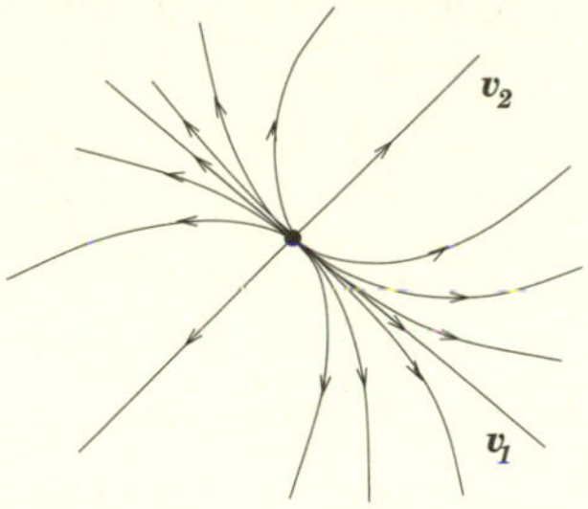

Fig. 4.9. Typical phase diagram for an unstable node

Example 4.5 Sketch the phase diagram for the system

$$\dot{\boldsymbol{x}} = A\boldsymbol{x} \quad \text{with } A = \begin{pmatrix} -2 & 1 \\ 1 & -2 \end{pmatrix},$$

and hence determine the graphs of x and y as functions of t where x and y are the solutions satisfying $x(0) = 1$, $y(0) = 0$. Here x, y are the coordinates of the vector $\boldsymbol{x}$, as in (4.13).

The eigenvalues λ of A are -3 or -1. It follows that $(0,0)$ is a stable node. The eigenvector for -3 is $\begin{pmatrix} -1 \\ 1 \end{pmatrix}$ and the eigenvector for -1 is $\begin{pmatrix} 1 \\ 1 \end{pmatrix}$. The phase diagram is as shown in Fig. 4.10, which also includes graphs of solutions satisfying $x(0) = 1, y(0) = 0$ (i.e., solutions corresponding to trajectory through $(1,0)$). □

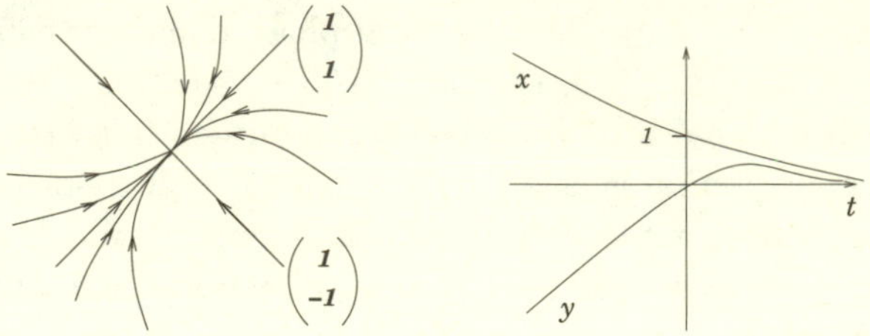

Fig. 4.10. Phase diagram and sketch of solution for Example 4.5

(c) **Distinct, real eigenvalues of opposite signs**: $\lambda_1 < 0 < \lambda_2$. The general solution is again given by (4.15). If $c_1 = 0$ and $c_2 \neq 0$, the trajectory is outwards along the ray in the direction of $c_2\boldsymbol{v}_2$. If $c_2 = 0$ and $c_1 \neq 0$, the trajectory is inwards along the ray in the direction of $c_1\boldsymbol{v}_1$. More generally, when $t \to \infty$ the solution is dominated by $c_2 e^{\lambda_2 t}\boldsymbol{v}_2$, and, when $t \to -\infty$ the solution is dominated by $c_1 e^{\lambda_1 t}\boldsymbol{v}_1$. The phase diagram is shown in Fig. 4.11. The point $(0,0)$ is called a *saddle point*.

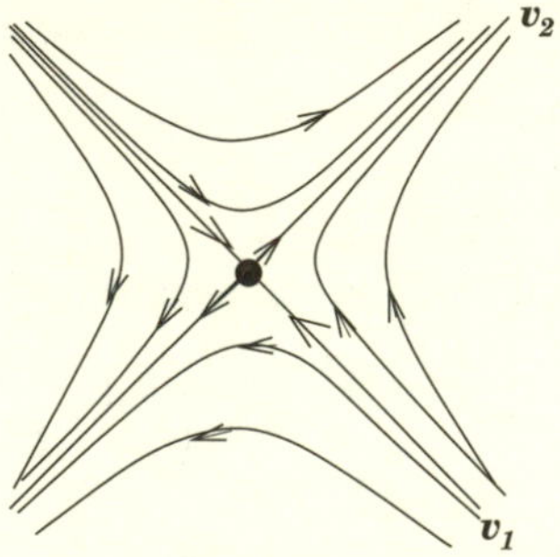

Fig. 4.11. Phase diagram of a saddle

Example 4.6 Sketch the phase diagram of

$$\dot{\boldsymbol{x}} = A\boldsymbol{x}, \quad \text{with } A = \begin{pmatrix} 1 & 1 \\ 4 & 1 \end{pmatrix}.$$

The eigenvalues of A are -1 and $= 3$. Hence $(0,0)$ is a saddle point. The eigenvector corresponding to -1 is $\begin{pmatrix} 1 \\ -2 \end{pmatrix}$ and the eigenvector corresponding to 3 is $\begin{pmatrix} 1 \\ 2 \end{pmatrix}$. Thus we obtain the phase diagram in Fig. 4.12.

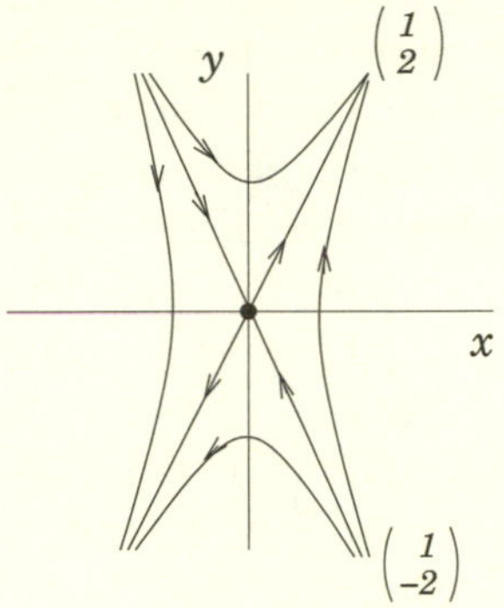

Fig. 4.12. Phase diagram for Example 4.6

(d) **One real, negative eigenvalue**: $\lambda_1 = \lambda_2 = \lambda < 0$.
(i) If there are two linearly independent eigenvectors $\boldsymbol{v}_1$ and $\boldsymbol{v}_2$ corresponding to λ, then the general solution is

$$\boldsymbol{x}(t) = e^{\lambda t}(c_1\boldsymbol{v}_1 + c_2\boldsymbol{v}_2).$$

Trajectories are in-going rays, and the phase diagram is shown in Fig. 4.13. The equilibrium point $(0, 0)$ is called a *stable star.*

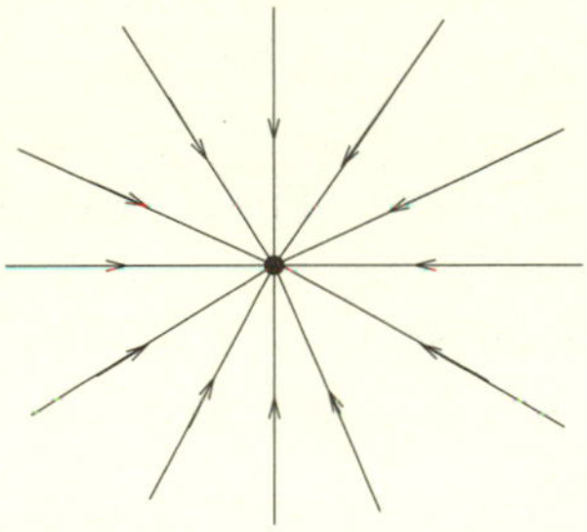

Fig. 4.13. Phase diagram for a stable star

(ii) Suppose that there is only one eigenvector $\boldsymbol{v}$ for the eigenvalue

λ. Then general solution is $\boldsymbol{x}(t) = c_1 e^{\lambda t}\boldsymbol{v} + c_2 e^{\lambda t}(t\boldsymbol{v} + \boldsymbol{w})$, where $\boldsymbol{w}$ satisfies $(A - \lambda I)\boldsymbol{w} = \boldsymbol{v}$. If $c_2 = 0$, we have a solution with a trajectory along $c_1\boldsymbol{v}$ with direction towards the origin. The general solution is dominated by $c_2 e^{\lambda t} t\boldsymbol{v}$ *both* as $t \to \infty$ *and* as $t \to -\infty$. All trajectories approach $(0,0)$ as $t \to \infty$ in the direction of $c_2 e^{\lambda t} t\boldsymbol{v}$ but, as $t \to -\infty$ they point in the opposite direction (since t is now negative). There are two possible 'S-shapes' of trajectories compatible with these observations, both shown in Fig. 4.14. The simplest way to work out the correct one is to go back to the original differential equations and consider $\dot{x} = ax + by$. If $b > 0$, then $\dot{x} > 0$ for $x = 0, y > 0$, so the trajectory crosses the positive y-axis in the direction of increasing x and we must have the left S-like picture. If $b < 0$, then $\dot{x} < 0$ for $x = 0, y > 0$, and we must have the right S-like picture. The point $(0,0)$ is called a *stable improper node* in both cases.

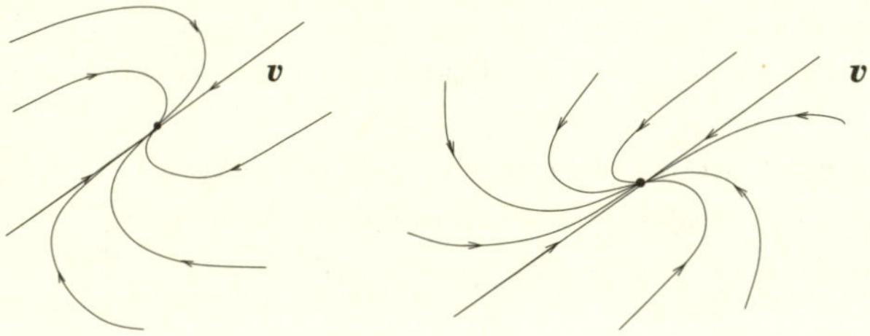

Fig. 4.14. Two possible phase diagrams for an improper node with given eigenvector $\boldsymbol{v}$

(e) **One real, positive eigenvalue**: $\lambda_1 = \lambda_2 = \lambda > 0$.
We obtain a similar phase diagram to case (d), but with $|\boldsymbol{x}(t)| \to \infty$ as $t \to \infty$ and $\boldsymbol{x}(t) \to (0,0)$ as $t \to -\infty$. We find that $(0,0)$ is an *unstable star* if there are two linearly independent eigenvectors and an *unstable improper node* if there is only one.

(f) **Complex eigenvalues**: $\lambda = \alpha + i\beta, \lambda^* = \alpha - i\beta$, $\beta \neq 0$.
As explained in Section 2.4.2, the general solution of (4.11) in this case can be written as

$$\boldsymbol{x}(t) = A\,\mathrm{Re}(e^{\lambda t}\boldsymbol{v}) + B\,\mathrm{Im}(e^{\lambda t}\boldsymbol{v}), \tag{4.16}$$

where $\boldsymbol{v}$ is the (complex) eigenvector for the eigenvalue λ. For the purpose of sketching, it is more convenient to combine the two arbitrary real constants A and B into the complex number $C = A - iB$ and write (4.16) as

$$\boldsymbol{x}(t) = \mathrm{Re}(Ce^{\lambda t}\boldsymbol{v}),$$

or, in the parametrisation $C = \rho e^{-i\phi}$, as

$$\boldsymbol{x}(t) = \rho e^{\alpha t}\mathrm{Re}(e^{i(\beta t - \phi)}\boldsymbol{v}).$$

Introducing real vectors $\boldsymbol{v}^1, \boldsymbol{v}^2$ so that $\boldsymbol{v} = \boldsymbol{v}^1 + i\boldsymbol{v}^2$ and expanding $e^{i(\beta t - \phi)} = \cos(\beta t - \phi) + i\sin(\beta t - \phi)$ we arrive at the expression

$$\boldsymbol{x}(t) = \rho e^{\alpha t}(\cos((\beta t - \phi)\boldsymbol{v}^1 - \sin(\beta t - \phi)\boldsymbol{v}^2). \tag{4.17}$$

To sketch, we distinguish three cases.
(i) $\alpha < 0$. Defining $T = \frac{2\pi}{|\beta|}$ we observe that

$$\boldsymbol{x}(t + T) = e^{\alpha T}\boldsymbol{x}(t),$$

i.e., at time $t + T$ the solution $\boldsymbol{x}$ points in the same direction as at time t, but the separation from the origin has decreased by the factor $e^{\alpha T}$. Therefore, all trajectories form inward spirals which tend to the stable equilibrium point $(0, 0)$ for $t \to \infty$. The point $(0, 0)$ is called a *stable spiral point* in this case.
(ii) $\alpha > 0$. Going through the same analysis as in (i) we conclude that the trajectories are outward spirals with $|\boldsymbol{x}(t)| \to \infty$ as $t \to \infty$. The equilibrium point $(0, 0)$ is unstable, and called *unstable spiral point* in this case.
(iii) $\alpha = 0$. The trajectories are all circles. The equilibrium point $(0, 0)$ is stable. However, since trajectories stay at a fixed distance from the equilibrium point, it is not asymptotically stable. It is called a *centre* in this case.

Phase diagrams for spirals and a centre are shown in Fig. 4.15. To work out if the orientation is clockwise or anticlockwise in each of these cases, we look at the differential equation in the original form (4.14), and work out the sign of $\dot{x}$ on the y-axis, e.g., at $(x, y) = (0, 1)$. Clearly, at this point $\dot{x} = b$. Hence, if $b > 0$, x increases when a trajectory crosses the y-axis. That means that the flow is from left to right in the upper half-plane, and so the flow is clockwise. Similarly, if $b < 0$ then the flow is anticlockwise. We do not need to worry about b vanishing: you should be able to show that $b \neq 0$ for any matrix of the form (4.13) whose eigenvalues have a non-vanishing imaginary part.

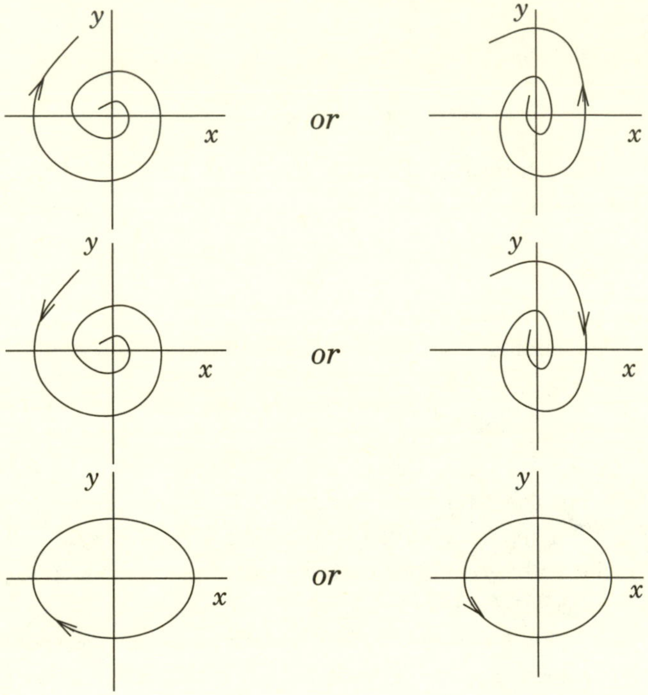

Fig. 4.15. Phase diagrams for (i) an outwards spiral (top), (ii) an inwards spiral and (iii) a centre

Example 4.7 Sketch the phase diagram of

$$\dot{\boldsymbol{x}} = A\boldsymbol{x} \quad \text{with } A = \begin{pmatrix} 3 & -1 \\ 5 & -1 \end{pmatrix}.$$

The eigenvalues of A are $\frac{2\pm\sqrt{4-8}}{2} = 1 \pm i$. Hence $(0,0)$ is an unstable spiral point. Note that the orientation of the spiral is determined by the fact that $\dot{x} < 0$ on the positive y-axis (at $(0,1)$ for example). This information is sufficient in most applications and does not require the eigenvectors. Computing the eigenvector

$\boldsymbol{v}$ for eigenvalue $1+i$ as

$$\boldsymbol{v} = \begin{pmatrix} 1 \\ 2-i \end{pmatrix}, \quad \text{so } \boldsymbol{v}^1 = \mathrm{Re}(\boldsymbol{v}) = \begin{pmatrix} 1 \\ 2 \end{pmatrix}, \boldsymbol{v}_2 = \mathrm{Im}(\boldsymbol{v}) = \begin{pmatrix} 0 \\ -1 \end{pmatrix},$$

allows one to deduce further details, like the direction of elongation of the spiral (can you figure out how?). The phase diagram is shown in Fig. 4.16. □

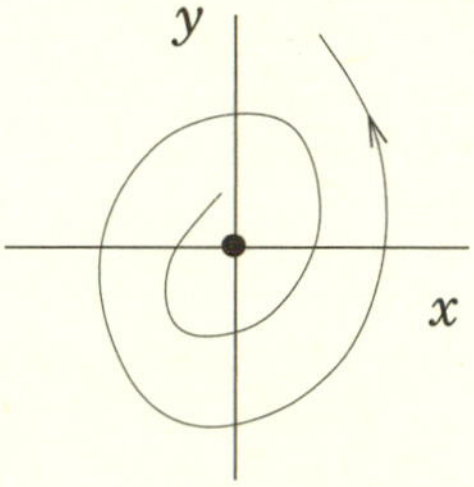

Fig. 4.16. Phase diagram for Example 4.7

(g) **One eigenvalue vanishes**: $\lambda_1 = 0$.
If $\boldsymbol{v}_1$ is the eigenvector corresponding to λ_1, then, by definition, $A\boldsymbol{v}_1 = 0$. Hence points on the line $L_1 = \{s\boldsymbol{v}^1 | s \in \mathbb{R}\}$ are equilibrium points. The phase diagram depends on whether the matrix A has one or two eigenvectors. If A has another eigenvalue $\lambda_2 \neq 0$ the general solution takes the form

$$\boldsymbol{x}(t) = c_1\boldsymbol{v}^1 + c_2 e^{\lambda_2 t}\boldsymbol{v}^2.$$

As t varies, the trajectories are half-lines in the direction of $c_2\boldsymbol{v}^2$, starting or ending at points on L_1. The direction of the flow depends on the sign on λ_2. If $\lambda_2 < 0$, the flow is towards the line L_1 which therefore consists of stable equilibrium points. If $\lambda_2 > 0$ the flow is away from the line L_1 which therefore consists of unstable equilibrium points. If 0 is a repeated eigenvalue, but A only has one eigenvector, the general solution takes the form

$$\boldsymbol{x}(t) = c_1\boldsymbol{v} + c_2(t\boldsymbol{v} + \boldsymbol{w}), \tag{4.18}$$

so that the trajectories are lines in the direction $c_2\boldsymbol{v}$, passing through $c_1\boldsymbol{v} + c_2\boldsymbol{w}$. All points on the line $L = \{s\boldsymbol{v} | s \in \mathbb{R}\}$ are equilibrium points, which are all unstable.

Another way of finding the trajectories in the case where A has 0 as an eigenvalue is to use the fact that the *rows of A must be multiples of each other.* This makes it possible to find equations of the trajectories in the form $y = ax + b$, as we shall now illustrate.

Example 4.8 Sketch the phase diagram for the system

$$\dot{\boldsymbol{x}} = A\boldsymbol{x}, \quad \text{with } A = \begin{pmatrix} 1 & -1 \\ 2 & -2 \end{pmatrix}.$$

The matrix A has determinant 0, and so 0 is an eigenvalue of A. The other eigenvalue is -1. A point (x, y) is an equilibrium point iff $x - y = 0$, $2x - 2y = 0$, i.e., if and only if $x = y$. Thus $x = y$ is a line of equilibrium points. Also, $y = y(x)$ is a trajectory for the system iff

$$\frac{dy}{dx} = \frac{\frac{dy}{dt}}{\frac{dx}{dt}} = \frac{2x - 2y}{x - y} = 2,$$

i.e., iff $y = 2x + c$. Hence we have the phase diagram shown in Fig. 4.17. The direction of the arrows can be deduced from the observation that $\dot{x} < 0$ when $y > x$ and $\dot{x} > 0$ when $y < x$. □

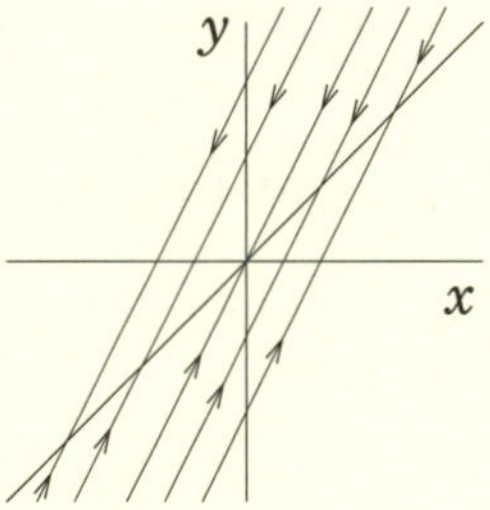

Fig. 4.17. Phase diagram for Example 4.8

Finally we note that, if there are two independent eigenvectors for the eigenvalue 0, the matrix A is necessarily the zero matrix. In this case, the equation (4.11) is simply $\dot{\boldsymbol{x}} = 0$: all points in the plane are stable equilibrium points. The findings of this section are summarised in Table 4.1.

Table 4.1. *Types of equilibrium points of linear systems $\dot{\boldsymbol{x}} = A\boldsymbol{x}$ and their stability properties*

	Eigenvalue	Type	Stability
(a)	$\lambda_1 < \lambda_2 < 0$	stable node	asym. stable
(b)	$0 < \lambda_1 < \lambda_2$	unstable node	unstable
(c)	$\lambda_1 = \lambda_2 < 0$	stable improper node or stable star	asym. stable
	$\lambda_1 = \lambda_2 > 0$	unstable improper node unstable star	unstable
(d)	$\lambda_1 < 0 < \lambda_2$	saddle point	unstable
(e)	$\lambda_{1,2} = \alpha \pm i\beta$		
	$\alpha < 0$	stable spiral point	asym. stable
	$\alpha > 0$	unstable spiral point	unstable
(f)	$\lambda_{1,2} = \pm i\beta$	centre	stable
(g)	$\lambda_1 = 0$		
	$\lambda_2 < 0$	stable line	asym. stable
	$\lambda_2 > 0$	unstable line	unstable
	$\lambda_2 = 0$	stable plane	stable

Exercise 4.5 Write the equation for the unforced but damped oscillating spring $m\ddot{x} + r\dot{x} + kx = 0$ as a second order system and analyse the nature and stability of the equilibrium point $(0, 0)$ as the real parameters m, r and k vary. Refer to the discussion in Section 3.2.2 and sketch phase diagrams for an overdamped, a critically damped and an underdamped spring.

Exercise 4.6 Consider the following two-dimensional systems of the form $\dot{\boldsymbol{x}} = A\boldsymbol{x}$, with A given by

$$(a) \begin{pmatrix} 3 & -2 \\ 2 & -2 \end{pmatrix} \quad (b) \begin{pmatrix} 5 & -1 \\ 3 & 1 \end{pmatrix} \quad (c) \begin{pmatrix} 1 & -5 \\ 1 & -3 \end{pmatrix}$$

$$(d) \begin{pmatrix} 2 & -5 \\ 1 & 2 \end{pmatrix} \quad (e) \begin{pmatrix} 3 & 2 \\ -2 & -1 \end{pmatrix}.$$

In each case,
(i) determine the type of the equilibrium point $(0, 0)$ and state whether it is stable, asymptotically stable or unstable,
(ii) sketch the phase diagram for the system,
(iii) sketch graphs of x and y against t for a solution $\boldsymbol{x}(t)$ passing through the point $(1, 0)$ at $t = 0$.

4.2 Nonlinear systems

4.2.1 The Linearisation Theorem

As promised, we now look at the linear approximation mentioned in Section 4.1.3 in more detail. We focus on autonomous systems of the form (4.12), which we write more explicitly in terms of the coordinates x, y of $\boldsymbol{x}$ as

$$\begin{aligned} \dot{x} &= f(x, y) \\ \dot{y} &= g(x, y). \end{aligned} \tag{4.19}$$

Here $f, g : \mathbb{R}^2 \to \mathbb{R}$ are continuously differentiable functions and we assume that $\boldsymbol{x}^* = (x^*, y^*)$ is an equilibrium point of (4.19) i.e., that $f(x^*, y^*) = g(x^*, y^*) = 0$. The linear approximation to (4.19) for $\boldsymbol{x}$ close to $\boldsymbol{x}^*$ is obtained by writing

$$x = x^* + \xi, \quad \text{and} \quad y = y^* + \eta, \tag{4.20}$$

and carrying out a Taylor expansion:

$$\begin{aligned} f(\boldsymbol{x}) &= f(\boldsymbol{x}^*) + \xi \frac{\partial f}{\partial x}(\boldsymbol{x}^*) + \eta \frac{\partial f}{\partial y}(\boldsymbol{x}^*) \\ &+ \text{higher powers in } \xi, \eta. \\ &\approx \xi \frac{\partial f}{\partial x}(\boldsymbol{x}^*) + \eta \frac{\partial f}{\partial y}(\boldsymbol{x}^*), \quad \text{if } \xi, \eta \text{ are small.} \end{aligned}$$

Similarly

$$g(\boldsymbol{x}) \approx \xi \frac{\partial g}{\partial x}(\boldsymbol{x}^*) + \eta \frac{\partial g}{\partial y}(\boldsymbol{x}^*), \quad \text{if } \xi, \eta \text{ are small.}$$

Noting from (4.20) that $\dot{x} = \dot{\xi}$ and $\dot{y} = \dot{\eta}$, we deduce that, for $\boldsymbol{x}$ close to $\boldsymbol{x}^*$, it is plausible that the nonlinear system (4.19) can be approximated by the linear system

$$\begin{aligned} \dot{\xi} &= \frac{\partial f}{\partial x}(\boldsymbol{x}^*)\xi + \frac{\partial f}{\partial y}(\boldsymbol{x}^*)\eta \\ \dot{\eta} &= \frac{\partial g}{\partial x}(\boldsymbol{x}^*)\xi + \frac{\partial g}{\partial y}(\boldsymbol{x}^*)\eta. \end{aligned} \tag{4.21}$$

This linear system is called the linearised equation of (4.19). It can be written in matrix form as

$$\frac{d}{dt}\begin{pmatrix} \xi \\ \eta \end{pmatrix} = A \begin{pmatrix} \xi \\ \eta \end{pmatrix},$$

where A is the Jacobian matrix at $\boldsymbol{x}^*$:

$$A = \begin{pmatrix} \frac{\partial f}{\partial x}(\boldsymbol{x}^*) & \frac{\partial f}{\partial y}(\boldsymbol{x}^*) \\ \frac{\partial g}{\partial x}(\boldsymbol{x}^*) & \frac{\partial g}{\partial y}(\boldsymbol{x}^*) \end{pmatrix}. \tag{4.22}$$

We would like to know if solutions of (4.19) and (4.21) behave similarly, at least close to $\boldsymbol{x}^*$. In fact the following can be proved.

Theorem 4.1 (Linearisation Theorem) *Let λ and μ be eigenvalues of the Jacobian matrix* (4.22). *If $Re(\lambda) \neq 0$, $Re(\mu) \neq 0$ and $\lambda \neq \mu$, then the qualitative behaviour of solutions of the nonlinear system* (4.19) *near $\boldsymbol{x}^* = (x^*, y^*)$ is the same as the behaviour of the linear system* (4.21) *near the equilibrium point* (0, 0).

The notion of 'qualitative behaviour near a point' can be made mathematically precise by using concepts from differential topology. The precise version of the theorem is called the Hartman–Grobman theorem. It was formulated and proved by D. M. Grobman and P. Hartman in 1959/60. Both the precise statement of the theorem and its proof are outside the scope of this book. However, we will try to clarify the claim of the theorem through some comments and then show how the theorem can be applied in a worked example.

In the terminology of Table 4.1, the equilibrium points to which the theorem applies are stable/unstable nodes, stable/unstable spiral points and saddle points. In the other cases, one can say the following.

(i) If $\lambda = i\omega$, then $\boldsymbol{x}^*$ is a centre for the linear system (4.21). It may remain a centre or become a stable or unstable spiral for the nonlinear system (4.19).

(ii) If $\lambda = \mu$ with $\lambda, \mu < 0$, then $\boldsymbol{x}^*$ is a stable improper node or stable star for the linear system (4.21), but it may be the same for the nonlinear system (4.19) or a stable node or stable spiral point.

(iii) If $\lambda = \mu$ with $\lambda, \mu > 0$, then $\boldsymbol{x}^*$ is an unstable improper node or unstable star for the linear system (4.21), but may be the same or an unstable node or an unstable spiral point for the nonlinear system (4.19).

Example 4.9 Find the equilibrium points and determine their nature for the system

$$\dot{x} = 2y + xy, \quad \dot{y} = x + y. \tag{4.23}$$

Hence plot a possible phase diagram.

The point (x, y) is an equilibrium point iff

$$2y + xy = 0 \quad \text{and} \quad x + y = 0.$$

The first equation is solved by $x = -2$ or $y = 0$, and the second requires $x = -y$. Hence, the equilibrium points are $(0, 0)$ and $(-2, 2)$. With $f(x, y) = 2y + xy$, we have $\frac{\partial f}{\partial x}(x, y) = y$ and $\frac{\partial f}{\partial y}(x, y) = 2 + x$. Similarly, with $g(x, y) = x + y$, we have $\frac{\partial g}{\partial x}(x, y) = 1$ and $\frac{\partial g}{\partial y}(x, y) = 1$. Hence, the linearised equation at $(0, 0)$ is

$$\dot{\boldsymbol{x}} = A\boldsymbol{x} \quad \text{with} \quad A = \begin{pmatrix} 0 & 2 \\ 1 & 1 \end{pmatrix}. \tag{4.24}$$

The eigenvalues of A are $\lambda_1 = -1$ and $\lambda_2 = 2$ with eigenvectors $\begin{pmatrix} -2 \\ 1 \end{pmatrix}$ and $\begin{pmatrix} 1 \\ 1 \end{pmatrix}$. Hence $(0, 0)$ is a saddle point for the linear system (4.24) and, by the Linearisation Theorem, also a saddle point for the nonlinear system (4.23). The linearised equation at the second equilibrium point $(-2, 2)$ is

$$\dot{\boldsymbol{x}} = A\boldsymbol{x}, \quad \text{with} \quad A = \begin{pmatrix} 2 & 0 \\ 1 & 1 \end{pmatrix}. \tag{4.25}$$

The matrix A has eigenvalues $\lambda_1 = 1$ and $\lambda_2 = 2$, with corresponding eigenvectors $\begin{pmatrix} 0 \\ 1 \end{pmatrix}$ and $\begin{pmatrix} 1 \\ 1 \end{pmatrix}$. Thus, the point $(-2, 2)$ is an unstable node for the linearised system and therefore also for the original nonlinear system (4.23). A possible phase diagram for (4.23) is shown in Fig. 4.18.

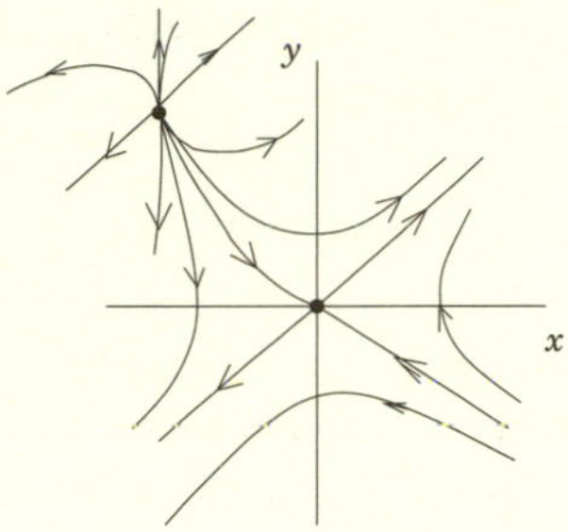

Fig. 4.18. Phase diagram for the system (4.23)

□

Exercise 4.7 Find all equilibrium points, determine their nature and sketch the phase plane for
(i) $\dot{x} = 1 - xy$, $\dot{y} = (x - 1)y$,
(ii) $\ddot{x} + x - x^3 = 0$.

Exercise 4.8 Consider the system

$$\dot{x} = y + x(1 - x^2 - y^2), \quad \dot{y} = -x + y(1 - x^2 - y^2). \tag{4.26}$$

(i) Show, using Exercise 2.5 or otherwise, that the system (4.26) becomes $\dot{r} = r(1 - r^2)$, $\dot{\phi} = 1$ in terms of the polar coordinates $x = r\cos\phi$, $y = r\sin\phi$.
(ii) Sketch the direction field for $\dot{r} = r(1 - r^2)$ in the (t, r)-half-plane; also sketch graphs $(t, r(t))$ of typical solutions.
(iii) Sketch the phase diagram of (4.26) and typical solutions.

4.2.2 Lyapunov functions

When the Linearisation Theorem does not apply, one needs to resort to other methods for establishing the stability of equilibrium points. In this final section, we describe a way of doing this via Lyapunov functions, named after Aleksandr Mikhailovich Lyapunov (1857–1918). To motivate this method, consider the non-linear system

$$\begin{aligned} \dot{x} &= \quad y - x\sqrt{x^2+y^2} \\ \dot{y} &= -x - y\sqrt{x^2+y^2}, \end{aligned} \tag{4.27}$$

which we studied already as one of the examples for checking the conditions of the Picard–Lindelöf Theorem in Example 2.2. The linearised equation at $(0,0)$ is

$$\dot{\boldsymbol{x}} = A\boldsymbol{x}, \quad \text{with} \quad A = \begin{pmatrix} 0 & 1 \\ -1 & 0 \end{pmatrix}.$$

The matrix A has eigenvalues $\pm i$. Hence $(0,0)$ is a centre for the linearised equation and the Linearisation Theorem does not allow us to draw any conclusions about the nonlinear system. For any solution of (4.27),

$$\begin{aligned} \dot{x}x &= \quad xy - x^2\sqrt{x^2+y^2} \\ \dot{y}y &= -xy - y^2\sqrt{x^2+y^2}, \end{aligned}$$

and hence

$$\frac{d}{dt}(x^2+y^2) = 2\dot{x}x + 2\dot{y}y = -2(x^2+y^2)^{\frac{3}{2}} < 0, \tag{4.28}$$

so that the combination x^2+y^2 is a strictly decreasing function of t. Since x^2+y^2 measures the separation from the origin, we deduce that $\boldsymbol{x}(t)$ gets closer to $(0,0)$ as t increases. Thus we deduce that $(0,0)$ is a stable equilibrium point of (4.27). In this particular example we can go further and integrate the ODE (4.28) for x^2+y^2 by separating variables. We find

$$x^2(t) + y^2(t) = \frac{1}{(t+c)^2},$$

which actually tends to 0 as $t \to \infty$.

The above technique is often useful in establishing whether an equilibrium point is stable or not. The function $x^2 + y^2$ can be replaced by any function V with similar properties.

Theorem 4.2 *Consider an autonomous system $\dot{\boldsymbol{x}} = \boldsymbol{f}(\boldsymbol{x})$ with an isolated equilibrium point $\boldsymbol{x}^*$. Let U be an open set containing $\boldsymbol{x}^*$ and let $V : U \to \mathbb{R}$ be a smooth function such that*

(1) $V(\boldsymbol{x}^*) = 0$ and $V(\boldsymbol{x}) > 0$ for all $\boldsymbol{x} \in U \setminus \{\boldsymbol{x}^*\}$,

(2) For every solution $\boldsymbol{x}$ of $\dot{\boldsymbol{x}} = \boldsymbol{f}(\boldsymbol{x})$ taking values in U,
$$\frac{d}{dt}(V(\boldsymbol{x}(t))) \leq 0.$$

Then $\boldsymbol{x}^*$ is a stable equilibrium point.

The requirement that V is 'smooth' can be made precise; see Jordan and Smith [3]. It is satisfied, for example, if V has a convergent Taylor series around $\boldsymbol{x}^*$ in U. However, the theorem still holds for functions which are only continuous and continuously differentiable in $U \setminus \{\boldsymbol{x}^*\}$, provided they satisfy additional technical requirements, also discussed in Jordan and Smith [3]. We sketch the idea of the proof for two-dimensional systems, referring to their book for details.

By assumption, the minimum value of V in U is 0 and attained only at the point $\boldsymbol{x}^*$. For $c > 0$ sufficiently small, the curves with equations of the form $V(x, y) = c$ form a system of closed curves surrounding $\boldsymbol{x}^*$. These look typically as shown in Fig. 4.19 (and the technical requirement on V mentioned above is precisely designed to ensure this). When $c = 0$, the level curve reduces to the point $\boldsymbol{x}^*$, and as c increases the level curve moves away from $\boldsymbol{x}^*$. This is illustrated in Fig. 4.19.

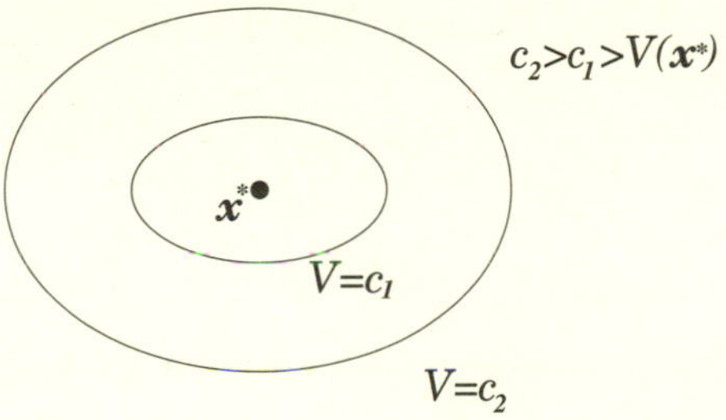

Fig. 4.19. Level curves around an equilibrium point

If $\boldsymbol{x}(0)$ lies inside the curve defined by $V(\boldsymbol{x}) = c$, i.e., if $V(\boldsymbol{x}(0)) < c$, then $V(\boldsymbol{x}(t)) < c$ for all $t > 0$ since $V(x(t))$ is decreasing by (2). Hence, the trajectory stays inside the curve defined by $V(\boldsymbol{x}) = c$ for all $t > 0$. It follows that $\boldsymbol{x}^*$ is a stable equilibrium point. □

Remarks

(i) If the condition (2) in the statement of the theorem is changed to $\frac{d}{dt}(V(\boldsymbol{x}(t))) > 0$ for every solution $\boldsymbol{x}(t)$, then the conclusion becomes $\boldsymbol{x}^*$ *is an unstable equilibrium point.*

(ii) If the condition (2) in the statement of the theorem is changed to $\frac{d}{dt}(V(\boldsymbol{x}(t))) < 0$ for every solution $\boldsymbol{x}(t)$, then the conclusion becomes $\boldsymbol{x}^*$ *is an asymptotically stable equilibrium point.* The proof of this result is quite involved - see again Jordan and Smith [3] for a discussion.

Functions V satisfying the requirements discussed above are called *Lyapunov functions.* They provide means of determining the stability of an equilibrium point when the Linearisation Theorem cannot be applied. In general, it is difficult to find a Lyapunov function for a given system of ODEs. It is sometimes possible to guess a Lyapunov function from geometrical or physical considerations, as in the energy discussion in Exercise 3.18. Also, if the equations of the trajectories can be found explicitly these may give Lyapunov functions.

Consider, for example, the equation $\ddot{x} + \sin(x) = 0$ governing pendulum motion and studied in Example 4.4. As we saw there, the trajectories are given by $V(x, y) = c$, where $V(x, y) = \frac{1}{2}y^2 - \cos x$. In physics, this equation expresses the conservation of the total energy V, which is made up of the kinetic term $\frac{1}{2}y^2$ and the potential term $-\cos x$. Clearly, V has minima at $(x, y) = (2\pi n, 0)$, $n \in \mathbb{Z}$. Furthermore

$$\begin{aligned}\frac{d}{dt}(V(x(t), y(t))) &= \frac{\partial V}{\partial x}\frac{dx}{dt} + \frac{\partial V}{\partial y}\frac{dy}{dt} \\ &= (\sin x)y + y(-\sin x) = 0.\end{aligned}$$

Hence $(2\pi n, 0)$ is a stable equilibrium point for any $n \in \mathbb{Z}$.

Exercise 4.9 Consider a population of predators (e.g., sharks) of size y, and and a population of prey (e.g., seals) of size x. Then x has positive growth rate when $y = 0$ (seals living happily in the absence of sharks) and $y > 0$ reduces the growth rate (sharks eating seals). On the other hand y has a negative growth rate when $x = 0$ (sharks starve) and $x > 0$ increases the growth rate of y (sharks enjoying seal dinners). The population dynamics is modelled by the following system of differential equations:

$$\dot{x} = ax - \alpha xy; \qquad \dot{y} = -by + \beta xy, \tag{4.29}$$

where a, b, α, β are positive real constants.
(i) Consider the function $V(x, y) = \beta x + \alpha y - b\ln(x) - a\ln(y)$ and show that, for any solution curve $(x(t), y(t))$ of (4.29) which does not pass through $(0, 0)$, $\frac{d}{dt}V(x(t), y(t)) = 0$.
(ii) Find the equilibrium points of (4.29). Determine the nature of the equilibrium points for the linearised system and deduce the nature of the equilibrium points of (4.29). Use the result of (a) in the case(s) where the linearisation theorem does not apply.
(iii) Draw a phase diagram of the predator-prey system. Pick values for the parameters a, b, α and β and sketch the populations $x(t)$ and $y(t)$ of prey and predators as functions of time. Interpret your sketches.

Exercise 4.10 The SIR model is a classic model in epidemiology, and a significant improvement on the simple model of Exercise 1.5. It considers three subpopulations, the susceptibles S, the infectives I and the removed individuals R (meaning immune or dead or in quarantine). The susceptibles can become infective, and the infectives can become removed, but no other transitions are considered. Diagrammatically,

$$S \to I \to R.$$

The total population $N = S + I + R$ remains constant. The model describes the movement between the classes by the system of differential equations

$$\frac{dS}{d\tau} = -\beta IS, \quad \frac{dI}{d\tau} = \beta IS - \gamma I, \quad \frac{dR}{d\tau} = \gamma I.$$

In terms of the fractions $x = S/N, y = I/N, z = R/N$, and the rescaled time variable $t = \gamma\tau$, the equations become

$$\frac{dx}{dt} = -R_0 xy, \qquad \frac{dy}{dt} = R_0 xy - y, \qquad \frac{dz}{dt} = y,$$

where $R_0 = \beta N/\gamma$ is an important parameter, called the reproductive ratio. The goal of this exercise is to analyse this model, and to extract quantitative predictions from it. Even though the model is relatively simple, its analysis requires care: the phase space is a bounded region in $\mathbb{R}^2$, and the equilibrium points lie on the boundary. You will therefore have to think carefully before applying the tools developed in this chapter.

(i) Explain why x, y, z all lie in the interval $[0, 1]$ and why $x + y + z = 1$. Since the equations for x and y do not involve z we can solve for x and y and then compute z via $z = 1 - x - y$. Mark the region $D = \{(x, y) \in \mathbb{R}^2 | 0 \leq x, y \leq 1, x + y \leq 1\}$ in the (x, y)-plane.

(ii) Consider the system

$$\frac{dx}{dt} = -R_0 xy, \quad \frac{dy}{dt} = R_0 xy - y. \tag{4.30}$$

Deduce a differential equation for y as a function of x and find its general solution.

(iii) Find all equilibrium points of (4.30) in the region D. Considering $R_0 \leq 1$ and $R_0 > 1$ separately, determine the stability properties of the equilbrium points. Draw a phase diagram for each of the cases $R_0 \leq 1$ and $R_0 > 1$.

(iv) Sketch the fractions y and x (infectives and susceptibles) as functions of t for trajectories starting near the disease-free state $(x, y) = (1, 0)$. Deduce that the number of infectives decreases if $R_0 \leq 1$ but that it initially increases if $R_0 > 1$. Also show that, regardless of the value of R_0, susceptibles remain in the population at the end of the disease.

5
Projects

This chapter contains five projects which develop the ideas and techniques introduced in this book. The first four projects are all based on recent research publications. Their purpose is to illustrate the variety of ways in which ODEs arise in contemporary research - ranging from engineering to differential geometry - and to provide an authentic opportunity for the reader to apply the techniques of the previous chapters. If possible, the projects could be tackled by a small group of students working as a team. The fifth project has a different flavour. Its purpose is to guide the reader through the proof of the Picard–Lindelöf Theorem. At the beginning of each project, I indicate the parts of the book which contain relevant background material.

5.1 Ants on polygons

(*Background: Chapters 1 and 2, Exercise 2.6*)

Do you remember the problem of four ants chasing each other at constant speed, studied in Exercise 2.6? We now look at two variations of this problem. In the first, we consider n ants, where $n = 2, 3, 4, 5 \ldots$, starting off on a regular n-gon. Here, a 2-gon is simply a line, a regular 3-gon an equilateral triangle, a 4-gon a square and so on. In the second, more difficult, variation we consider 4 ants starting their pursuit on a rectangle with side lengths in the ratio 1:2. This innocent-sounding generalisation turns out to be remarkably subtle and rich, and is the subject of recent research reported in Chapman, Lottes and Trefethen [4].

(i) Draw pictures of regular n-gons for $n = 2, 3, 4, 5, 6$ inscribed in a unit circle centred at the origin of the complex plane. Imagine an ant at each of the vertices of the n-gon and introduce complex numbers z_i, $i = 1 \ldots, n$ for the positions of the ants, numbered in anti-clockwise order. Mark the complex numbers in your pictures, which should be generalisations of Fig. 2.1. At time $t = 0$ the ants start chasing each other, walking at constant speed v, with the i-th ant always heading towards the $(i+1)$th ant, and the nth ant heading towards the first ant. Use the symmetry of the initial condition to reduce this problem to a single ODE for a complex-valued function, as in Exercise 2.6. Solve this ODE and compute the time it takes the ants to reach the centre. What happens as $n \to \infty$?

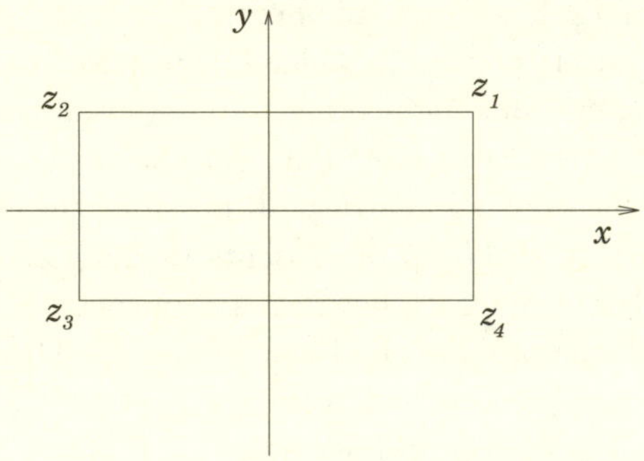

Fig. 5.1. Complex numbers as coordinates for the positions of four ants on a rectangle

(ii) Now consider four ants on the vertices of a rectangle, with side lengths 2 and 1, as shown in Fig. 5.1. The ants start chasing each other at $t = 0$, as in (i) above. Show, using the Picard–Lindelöf Theorem as in Exercise 2.6, that, at all times $t \geq 0$, $z_3(t) = -z_1(t)$ and $z_4(t) = -z_2(t)$, so that the ants' positions are always on the corners of a parallelogram. Deduce that the first two ants' position vectors satisfy the system of ODEs

$$\frac{dz_1}{dt} = \frac{z_2 - z_1}{|z_2 - z_1|}, \qquad \frac{dz_2}{dt} = \frac{-z_2 - z_1}{|z_1 + z_2|}. \tag{5.1}$$

Now set

$$\ell_1 e^{i\phi} = z_2 - z_1, \quad \ell_2 e^{i(\phi-\theta)} = z_2 + z_1 \tag{5.2}$$

so that $\ell_1 \geq 0$ and $\ell_2 \geq 0$ are the lengths of the sides of the parallelogram. Find the system of ODEs obeyed by ℓ_1, ℓ_2, θ and ϕ and show that the ants collide at $t = 3/2$. Finally, solve the system (5.1) numerically and draw pictures of the configurations as the ants chase each other. What do you notice as $t \to 3/2$? Chapman, Lottes and Trefethen [4] claim that the scale of the configuration has shrunk by a factor of $10^{427907250}$ after the first rotation. Can you reproduce this result? (Hard! See their paper for a solution.)

5.2 A boundary value problem in mathematical physics

(*Background: Chapters 1 and 2, Exercise 3.9*)

This project deals with a boundary value problem which arises in a mathematical model for vortices. If you are interested in the background to this problem have a look at Schroers [5]. The mathematical description of the vortex involves functions $\boldsymbol{\phi} : \mathbb{R}^2 \to \mathbb{R}^3$ and $\boldsymbol{A} : \mathbb{R}^2 \to \mathbb{R}^2$. The function $\boldsymbol{\phi}$ is constrained to satisfy $\phi_1^2 + \phi_2^2 + \phi_3^2 = 1$, and therefore takes values on the unit sphere. We use Cartesian coordinates (x_1, x_2) and polar coordinates (r, θ) in the plane, with the usual relation $x_1 = r\cos\theta$ and $x_2 = r\sin\theta$. We write

$$\boldsymbol{\phi} = \begin{pmatrix} \phi_1 \\ \phi_2 \\ \phi_3 \end{pmatrix}, \qquad \boldsymbol{A} = \begin{pmatrix} A_1 \\ A_2 \end{pmatrix}, \qquad \boldsymbol{n} = \begin{pmatrix} 0 \\ 0 \\ 1 \end{pmatrix},$$

and define

$$B = \frac{\partial A_2}{\partial x_1} - \frac{\partial A_1}{\partial x_2}, \qquad D_i\boldsymbol{\phi} = \frac{\partial \boldsymbol{\phi}}{\partial x_i} + A_i \boldsymbol{n} \times \boldsymbol{\phi},$$

with $i = 1, 2$. The system of PDEs we are interested in is

$$D_1\boldsymbol{\phi} = \boldsymbol{\phi} \times D_2\boldsymbol{\phi}, \qquad B = \boldsymbol{n}\cdot\boldsymbol{\phi} - 1. \tag{5.3}$$

We are looking for a solution of the form

$$\phi(r,\theta) = \begin{pmatrix} \sin f(r)\cos 2\theta \\ \sin f(r)\sin 2\theta \\ \cos f(r) \end{pmatrix},$$

$$A_1(r,\theta) = -\frac{2a(r)}{r}\sin\theta, \quad A_2(r) = \frac{2a(r)}{r}\cos\theta, \tag{5.4}$$

for real-valued functions f and a.

(i) Insert (5.4) into (5.3) to obtain the following ordinary differential equations for f and a:

$$f' = -2\frac{1+a}{r}\sin f, \qquad a' = \frac{r}{2}(\cos f - 1). \tag{5.5}$$

(ii) Solve (5.5) numerically subject to the boundary conditions

$$f(0) = \pi, \quad a(0) = 0 \quad \text{and} \quad \lim_{r\to\infty} f(r) = 0. \tag{5.6}$$

For each solution plot both f and a as functions of r. Also plot the function $B(r)$. *Hint*: if you do not know how to start this problem, have a look at Exercise 3.9, which introduces the idea of solving a boundary value problem using the shooting method.

5.3 What was the trouble with the Millennium Bridge?

(*Background: Chapters 2 and 3, particularly Section 3.2*)

The Millennium Bridge in London is a steel suspension footbridge over the Thames. Soon after the crowd streamed onto the bridge on the opening day in June 2000, the bridge began to sway from side to side. Pedestrians fell into step with the bridge's oscillations, inadvertently amplifying them. The bridge had to be closed for almost two years to allow for major re-engineering. In this project you will study a mathematical model for the bridge's oscillations, proposed in a short article [6] published in Nature. The mathematical model combines a standard, linear model of the bridge's lateral oscillation with a model of the pedestrians, and in particular of the way they synchronise their gait with the bridge's oscillations. Similar synchronisation phenomena or 'synchrony' are important in many physical and biological systems. You can find a lot of additional information about the Millennium Bridge on the internet. A good starting point is Abrams [7].

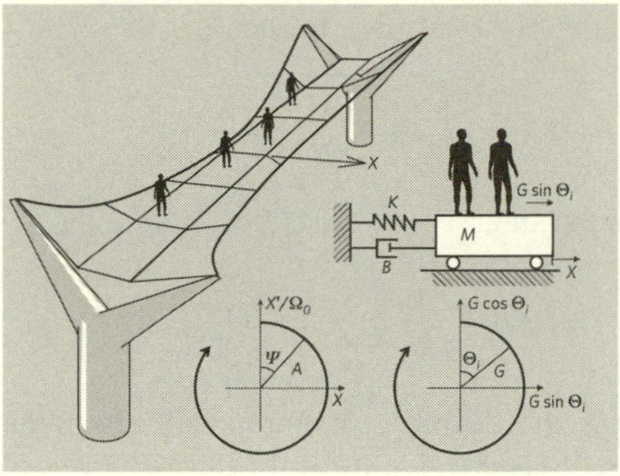

Fig. 5.2. Effect of pedestrian crowding on London's Millennium Bridge. The lateral mode of vibration of the bridge can be represented by the mass-spring-damper system shown on the top right. See main text for the definition of the variables. Figure reproduced from Strogatz et al. [6] with permission of the Nature Publishing Group

The mathematical model in Strogatz et al. [6] consists of a real function $X(t)$ for the bridge's lateral displacement from the equilibrium position, and angular coordinates Θ_i, $i = 1, \ldots, N$ to model the pedestrians: the angle Θ_i increases by 2π for a full left/right walking cycle by the i-th pedestrian, as illustrated in Fig. 5.2. The bridge's displacement is affected by the pedestrians according to

$$M\frac{d^2X}{dt^2} + B\frac{dX}{dt} + KX = G\sum_{i=1}^{N} \sin\Theta_i, \tag{5.7}$$

while the bridge's effect on the pedestrians' gait is described by

$$\frac{d\Theta_i}{dt} = \Omega_i + CA\sin(\Psi - \Theta_i + \alpha). \tag{5.8}$$

Here M, B, K, Ω_i, C and α are real constants, but A and Ψ are functions of t and related to the function X via

$$X = A\sin\Psi, \qquad \frac{dX}{dt} = A\sqrt{\frac{K}{M}}\cos\Psi. \tag{5.9}$$

You will need to solve the ODEs (5.7) and (5.8) numerically, with realistic values of N, M, B, K, Ω_i, C and α and initial conditions which you may choose. In order to interpret the results, Strogatz et al. [6] consider the 'wobble amplitude'

$$A = \sqrt{X^2 + \frac{M}{K}\left(\frac{dX}{dt}\right)^2}, \tag{5.10}$$

and the order parameter

$$R = \frac{1}{N}\sum_{j=1}^{N} \exp(i\Theta_j), \tag{5.11}$$

which measures the degree of coherence of the pedestrians' gaits (why?). You will need to compute both of these as functions of t.

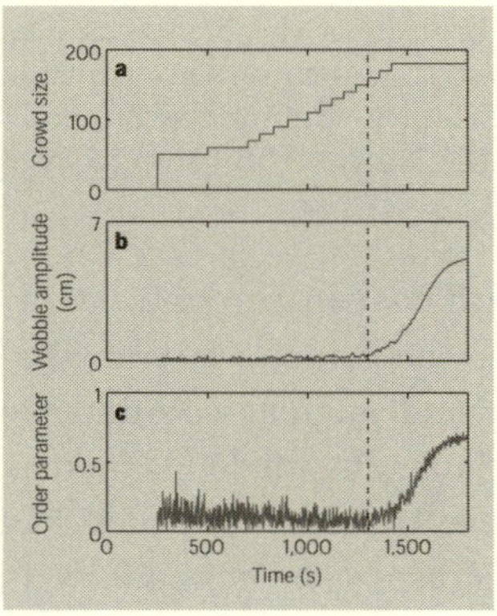

Fig. 5.3. (a) Number of pedestrians, N, increasing stepwise, as used in a diagnostic wobble test on the north span of the bridge in Dec. 2000. (b) Predicted amplitude A (5.10) of the bridge's lateral vibrations. (c) Predicted degree of phase coherence among the pedestrians, as measured by the order parameter R (5.11).
Reproduced from Strogatz et al. [6] with permission of the Nature Publishing Group

(i) Begin with a single pedestrian, so $N = 1$, and decouple the equations by setting $G = C = 0$. Use realistic parameter values for M, K and Ω_1 (use (5.13) as a guide) and pick three values for B corresponding, respectively, to underdamping, critical damping and overdamping of the bridge. In each case, pick initial conditions, solve the equations (5.7) for X and (5.8) for Θ_1 numerically and plot both X and Θ_1 as functions of t. Also plot the wobble amplitude (5.10) as a function of t. Discuss your findings.

(ii) Repeat the above, but now set $G = 0.3$ (so that the pedestrian's effect on the bridge is taken into account). Keep $C = 0$ (the bridge does not affect the pedestrian) and $N = 1$. Does the result make sense to you?
(iii) Now consider the effect of the bridge on the pedestrian and set $C = 16$. Then expand the right hand side of (5.8) using an appropriate trigonometric identity and the equation (5.9) to replace $A \sin \Psi$ and $A \cos \Psi$. Show that (5.8) becomes

$$\frac{d\Theta_i}{dt} = \Omega_i + C \left(X \cos(\alpha - \Theta_i) + \sqrt{\frac{M}{K}} \frac{dX}{dt} \sin(\alpha - \Theta_i) \right). \quad (5.12)$$

Now solve the coupled system (5.7) and (5.12), still keeping $N = 1$. Use several sets of parameter values for $M, B, K, \Omega_1, \alpha$. In each case, plot X, A and Θ_1 as functions of t.
(iv) Finally, consider a crowd of pedestrians on the Millennium Bridge: move up to $N = 10, N = 100$ and $N = 180$. In each case you will need to generate a random set of gait frequencies $\Omega_i, i = 1, \ldots, N$, with a realistic distribution and mean. You can take the following parameters for the north span of the bridge [8].

$$\begin{aligned} &M \approx 113 \times 10^3 \text{kg}, \\ &8.78 \times 10^3 \text{kg/s} \leq B \leq 11.7 \times 10^3 \text{kg/s}, \\ &K \approx 4730 \times 10^3 \text{kg/s}^2. \end{aligned} \quad (5.13)$$

Reasonable values for the remaining parameters are $G \approx 30$ Newton and $C \approx 16\ \text{m}^{-1}\text{s}^{-1}$. Now solve the system of ODEs for these parameter values numerically, with initial conditions of your choice. Can you reproduce the behaviour shown in Fig. 5.3?

5.4 A system of ODEs arising in differential geometry

(Background: Chapters 2 and 4)

In this project you will explore the non-linear system

$$\begin{aligned} \frac{da}{dr} &= \frac{1}{2rc}(a^2 - (b-c)^2) \\ \frac{db}{dr} &= \frac{b}{2acr}(b^2 - (a-c)^2) \\ \frac{dc}{dr} &= \frac{1}{2ra}(c^2 - (a-b)^2), \end{aligned} \quad (5.14)$$

of ordinary differential equations for real-valued functions a, b and c of one real variable r, which we considered briefly at the very beginning of this book, in Section 1.1.2. These equations arise in Riemannian geometry when one tries to find four-dimensional metrics with rotational symmetry and a special property, called self-duality. Further background on the origin of these equations in differential geometry and mathematical physics can be found in Atiyah and Hitchin [9].

(i) Show that the functions

$$a(r) = b(r) = r\sqrt{1+\frac{2}{r}}, \qquad c(r) = \frac{2}{\sqrt{1+\frac{2}{r}}} \tag{5.15}$$

satisfy the system of equations (5.14) on the interval $(0, \infty)$.

(ii) Using numerical methods, solve the system (5.14) on the interval $[\pi, \infty)$ subject to the initial conditions

$$a(\pi) = 0, \quad b(\pi) = \pi, \; c(\pi) = -\pi. \tag{5.16}$$

Hint: Since a vanishes at $r = \pi$ and $1/a$ appears on the right hand side of (5.14) you will need to find an approximate solution near $r = \pi$. To do this, let $x = r - \pi$, and try to find a solution of the form

$$a(r) = Ax^\alpha, \; b(r) = \pi + Bx^\beta, \; c(r) = -\pi + Cx^\gamma,$$

where you can determine $A, B, C, \alpha, \beta, \gamma$ by inserting into (5.14) and keeping only lowest powers in x (compare Exercise 3.9). Use this approximate solution to compute the values of a, b, c for some fixed value $r = r_0$ close to π. Then integrate (5.14) numerically out to some large value of r, with initial conditions imposed at r_0.

(iii) Express the system (5.14) in terms of the independent variable u which is related to r via

$$\frac{du}{dr} = -\frac{b^2}{rac}.$$

Show that, if a, b, c satisfy the resulting system of ODEs, then the ratios $x = \frac{a}{b}$ and $y = \frac{c}{b}$ satisfy the system

$$\begin{aligned} \frac{dx}{du} &= x(1-x)(1+x-y), \\ \frac{dy}{du} &= y(1-y)(1+y-x). \end{aligned} \tag{5.17}$$

Find all critical points of this system and determine their nature. Hence sketch a phase diagram of (5.17) and, in it, sketch the trajectories corresponding to the explicit solution (5.15) and the solution found numerically in (ii).

5.5 Proving the Picard–Lindelöf Theorem

(*Background: Chapters 1 and 2, some knowledge of real analysis*)
The following notes are designed as a guide through the proof of two versions of the Picard–Lindelöf Theorem. They could be used as the basis for self-study or for a seminar on this topic. You will need to fill in some gaps in the proof yourself, and you may need to look up some additional material on the internet or in books.

5.5.1 The Contraction Mapping Theorem

We will use the basic linear algebra revised in Section 2.3.2, but we also require tools from analysis and functional analysis. Recall that a *norm* on a real or complex vector space V is a map $||\cdot|| : V \to \mathbb{R}$ which satisfies

$$\begin{aligned} &||v|| \geq 0 \text{ for all } v \in V, \qquad ||v|| = 0 \Leftrightarrow v = 0, \\ &||\lambda v|| = |\lambda|\, ||v||, \text{ for } \lambda \in \mathbb{R} \text{ (or } \mathbb{C}), \; v \in V, \\ &||v + w|| \leq ||v|| + ||w||. \end{aligned}$$

A sequence $(v_n)_{n\in\mathbb{N}}$ of elements of V is called a *Cauchy sequence*, named after Augustin Louis Cauchy (1789–1857), if

$$\lim_{n,m\to\infty} ||v_n - v_m|| = 0.$$

We say that a subset S of a normed vector space is *complete* if every Cauchy sequence in S has a limit in S. Complete, normed vector spaces are named after Stefan Banach (1892–1945):

Definition 5.1 A Banach space is a normed, complete vector space over $\mathbb{R}$ or $\mathbb{C}$.

Exercise 5.1 Show that $||\boldsymbol{x}||_m = \max(|x_1|, \ldots, |x_n|)$ and $||\boldsymbol{x}||_e = \sqrt{x_1^2 + \ldots x_n^2}$ define norms on $\mathbb{R}^n$ and that, with either of these norms, $\mathbb{R}^n$ is a Banach space. You may assume the completeness of $\mathbb{R}$.

One can show that on $\mathbb{R}^n$ all norms are equivalent; in particular, this means that they define the same notions of continuity and convergence. In the following we write $||\cdot||$ for a norm on $\mathbb{R}^n$. Our arguments are valid regardless of the norm we choose.

Exercise 5.2 Let K be a compact subset of $\mathbb{R}$, for example, a closed and bounded interval. Show that the set $C(K, \mathbb{R}^n)$ of continuous functions from K to $\mathbb{R}^n$ with supremum norm

$$||\boldsymbol{x}||_\infty = \sup_{t \in K} ||\boldsymbol{x}(t)|| \tag{5.18}$$

is a Banach space.

Definition 5.2 A map $F : D \subset V \to V$ from a subset D of a normed space V to V is called a contraction map if there exists a $k < 1$ such that for all $v, w \in D$,

$$||F(v) - F(w)|| \le k||v - w||.$$

The following result provides the key to proving the Picard–Lindelöf Theorem.

Theorem 5.1 (Contraction Mapping Theorem) *Let Ω be a closed subset of a Banach space V and $\Gamma : \Omega \to \Omega$ a contraction map. Then there exists a unique $v^* \in \Omega$ so that*

$$\Gamma(v^*) = v^*. \tag{5.19}$$

The element v^* is called the fixed point of the contraction map Γ.

Before we look at the proof, we stress that there are *two* important assumptions in the theorem: (i) that the subset Ω is mapped to itself, and (ii) that Γ is a contraction map.

Exercise 5.3 Consider the Banach space $V = \mathbb{R}$. For each of the following maps, determine if it is a contraction map, and if it

satisfies the conditions of the Contraction Mapping Theorem.
(i) $f : \mathbb{R} \to \mathbb{R},\ f(x) = \frac{1}{2}x + 2$, (ii) $f : \mathbb{R} \to \mathbb{R},\ f(x) = x^2$,
(iii) $f : [0, \frac{1}{3}] \to [0, \frac{1}{3}],\ f(x) = x^2$.

The idea for the proof of the Contraction Mapping Theorem is to pick an arbitrary element $v_0 \in \Omega$ and to consider the sequence $(v_n)_{n\in\mathbb{N}}$ defined via $v_{n+1} = \Gamma(v_n)$. One shows that $(v_n)_{n\in\mathbb{N}}$ is a Cauchy sequence and, since Ω is a closed subset of a Banach space, that this Cauchy sequence has a limit $v^* \in \Omega$. Finally, one uses the uniqueness of the limit to deduce that $\Gamma(v^*) = v^*$.

Exercise 5.4 Consider again the Banach space $V = \mathbb{R}$. Consider the map $F : [0,1] \to [0,1],\ F(x) = e^{-x}$, and show that it is a contraction map. Sketch a graph of F together with a graph of the identity map $\mathrm{id}(x) = x$. Explain why the intersection of the two graphs is the fixed point x^* of F and draw a picture of the sequence $(x_n)_{n\in\mathbb{N}}$, defined via $x_{n+1} = F(x_n)$, starting with an arbitrary $x_0 \in [0,1]$. Give a geometrical argument why this sequence converges to x^* if F is a contraction map.

Exercise 5.5 Write up a careful proof of the Contraction Mapping Theorem.

5.5.2 Strategy of the proof

We fix a value t_0 where we impose our initial condition $\boldsymbol{x}(t_0) = \boldsymbol{x}_0$ and introduce some notation for intervals and neighbourhoods which we will use throughout this section. We write

$$I_T = [t_0 - T, t_0 + T],$$
$$B_d = \{\boldsymbol{y} \in \mathbb{R}^n \mid ||\boldsymbol{y} - \boldsymbol{x}_0|| \leq d\}, \tag{5.20}$$

and allow the parameters T and d to take the value ∞, with the convention that $I_\infty = \mathbb{R}$ and $B_\infty = \mathbb{R}^n$. As discussed in Exercise (1.13) (for the case $n = 1$), the initial value problem

$$\frac{d\boldsymbol{x}}{dt} = \boldsymbol{f}(t, \boldsymbol{x}), \qquad \boldsymbol{x}(t_0) = \boldsymbol{x}_0 \tag{5.21}$$

for a vector-valued differentiable function $\boldsymbol{x} : I_T \to \mathbb{R}^n$ is equivalent to the integral equation

$$\boldsymbol{x}(t) = \boldsymbol{x}_0 + \int_{t_0}^{t} \boldsymbol{f}(s, \boldsymbol{x}(s))\, ds.$$

The integral equation makes sense for any continuous function $\boldsymbol{x} : I_T \to \mathbb{R}^n$. The idea behind the proof of the Picard–Lindelöf Theorem is to show that, for a suitable $\delta > 0, d > 0$ and with the notation (5.20), the map

$$\begin{aligned} \Gamma &: C(I_\delta, B_d) \to C(I_\delta, B_d), \\ \Gamma(\boldsymbol{x})\,(t) &= \boldsymbol{x}_0 + \int_{t_0}^{t} f(s, \boldsymbol{x}(s))\, ds \end{aligned} \tag{5.22}$$

is well-defined and a contraction mapping. Its unique fixed point is then the unique solution of the initial value problem (5.21). In order to apply the Contraction Mapping Theorem, we need to

(i) choose δ and d so that Γ maps $C(I_\delta, B_d)$ to itself, and
(ii) impose suitable conditions to ensure that Γ is a contraction mapping.

There are different versions of the Picard–Lindelöf Theorem in which these requirements are met in different ways. However, all versions assume a condition on the function f, called the Lipschitz condition, named after Rudolf Lipschitz (1832–1903):

Definition 5.3 Let U be a subset of $\mathbb{R}^m$. A function $\boldsymbol{f} : U \to \mathbb{R}^n$ satisfies a Lipschitz condition if there exists a real, positive constant k so that $||\boldsymbol{f}(\boldsymbol{x}) - \boldsymbol{f}(\boldsymbol{y})|| \leq k||\boldsymbol{x} - \boldsymbol{y}||$ for all $\boldsymbol{x}, \boldsymbol{y} \in U$. In this case, we also say that the function $\boldsymbol{f}$ is Lipschitz continuous.

Exercise 5.6 (i) Which of the following functions are Lipschitz continuous?
(a) $f : \mathbb{R} \to \mathbb{R}$, $f(x) = x^2$, (b) $f : [1, 6] \to \mathbb{R}$, $f(x) = x^2$,
(c) $f : (0, 1) \to \mathbb{R}$, $f(x) = 1/(1 - x)$.
(ii) Show that Lipschitz continuous functions $f : \mathbb{R} \to \mathbb{R}$ are continuous. Show also that any continuously differentiable function $f : I \to \mathbb{R}$ on a closed and bounded interval $I \subset \mathbb{R}$ is Lipschitz continuous.

5.5.3 Completing the proof

We study the proofs of two versions of the Picard–Lindelöf Theorem:

(G) The global version: we assume $d = \infty$, and require $\boldsymbol{f}$ to satisfy a global Lipschitz condition. Then we can prove uniqueness and existence of the global solution.

(L) The local version: we allow d to be finite, and require only that $\boldsymbol{f}$ satisfies a Lipschitz condition in $I_T \times B_d$. Then we can prove uniqueness and existence of the solution in some interval containing t_0 (local solution).

Theorem 5.2 *(G) Let J be an arbitrary interval with $t_0 \in J$, or all of $\mathbb{R}$, and let $\boldsymbol{f} : J \times \mathbb{R}^n \to \mathbb{R}^n$ be a continuous function satisfying a Lipschitz condition in $\boldsymbol{x} \in \mathbb{R}^n$, with Lipschitz constant independent of t, i.e., there is a $k > 0$ such that*

$$\text{for all } t \in J,\ \boldsymbol{x}, \boldsymbol{y} \in \mathbb{R}^n \quad ||\boldsymbol{f}(t, \boldsymbol{x}) - \boldsymbol{f}(t, \boldsymbol{y})|| \le k||\boldsymbol{x} - \boldsymbol{y}||.$$

Then the initial value problem (5.21) *has a unique solution which is defined everywhere on J.*

Sketch of Proof Let $\tau > 0$ and consider the Banach space $C(I_\tau, \mathbb{R}^n)$ of continuous vector-valued function on I_τ, equipped with the supremum norm (5.18). With $\boldsymbol{x} \in C(I_\tau, \mathbb{R}^n)$ one checks that $\Gamma(\boldsymbol{x}) \in C(I_\tau, \mathbb{R}^n)$, where Γ is defined in (5.22). We need to show that

$$\Gamma : C(I_\tau, \mathbb{R}^n) \to C(I_\tau, \mathbb{R}^n)$$

is a contraction mapping. For any $\boldsymbol{x}_1, \boldsymbol{x}_2 \in C(I_\tau, \mathbb{R}^n)$ we use the Lipschitz continuity of f and standard inequalities to deduce

$$\begin{aligned} ||\Gamma(\boldsymbol{x}_1) - \Gamma(\boldsymbol{x}_2)||_\infty &= \sup_{t \in I_\tau} \left|\left| \int_{t_0}^t \boldsymbol{f}(s, \boldsymbol{x}_1(s)) - \boldsymbol{f}(s, \boldsymbol{x}_2(s))\, ds \right|\right| \\ &\le k \sup_{t \in I_\tau} \int_{t_0}^t ||\boldsymbol{x}_1(s) - \boldsymbol{x}_2(s)||\, ds \\ &\le \tau k ||\boldsymbol{x}_1 - \boldsymbol{x}_2||_\infty. \end{aligned} \tag{5.23}$$

Hence, Γ is indeed a contraction provided we pick $\tau < 1/k$. This guarantees the existence of a fixed point of Γ and hence of a solution $\boldsymbol{x}$ of (5.21) in the interval I_τ. This solution has some value

$x(t_0')$ at $t_0' = t_0 + \tau$, which we take as the initial value for a new initial value problem at t_0'. Repeating the arguments above we deduce the existence and uniqueness of the solution in the interval $[t_0, t_0 + 2\tau]$; by an analogous argument we show the existence and uniqueness in the interval $[t_0 - 2\tau, t_0]$. Iterating these steps establishes existence and uniqueness of a solution on the interval $[t_0 - l\tau, t_0 + m\tau]$ for any $l, m \in \mathbb{N}$, provided $[t_0 - l\tau, t_0 + m\tau] \subset J$. If $J = \mathbb{R}$ we deduce the existence and uniqueness of the solution for any $t \in \mathbb{R}$ by taking l or m large enough. If J is an open or closed interval, we adapt, if necessary, the step size τ as we approach any boundary to prove the existence and uniqueness of the solution for any $t \in J$. □

Exercise 5.7 Fill in the details in the proof of Theorem 5.2.

Exercise 5.8 Show that Theorem 5.2 guarantees the existence and uniqueness of the solution of any initial value problem for the linear system (2.19) if the $n \times n$ matrix-valued function A is constant.

The proof of Theorem 5.2 provides the simplest illustration of how the Contraction Mapping Theorem can be used to establish existence and uniqueness of solutions of an initial value problem. Unfortunately, the assumption of a global Lipschitz condition with a Lipschitz constant k independent of t is very strong, and many interesting examples do not satisfy it. However, the proof can quite easily be adapted to deal with more general situations, as the following exercise illustrates.

Exercise 5.9 Consider the general linear system (2.19) with an arbitrary $A \in C(\mathbb{R}, \mathbb{R}^{n^2})$. Show that (2.19) satisfies the assumptions of (5.2) on any closed and bounded interval J containing t_0. Hence prove the global existence and uniqueness theorem 2.2 for the general linear system (2.19). *Hint*: If you have trouble completing the proof, look at the case $n = 1$ first and then try to generalise.

We finally come to the local version of the Picard–Lindelöf Theorem, which is the one we used in this book. The reason why this version is most useful in practice is that, as illustrated in Exercise 5.6 for $n = 1$, many functions $f : \mathbb{R}^n \to \mathbb{R}^n$ satisfy a Lipschitz condition on a closed and bounded interval even if they are not Lipschitz continuous on all of $\mathbb{R}^n$.

Theorem 5.3 *(L) Let I_T and B_d be defined as in* (5.20), *with both T and d finite. Assume that $\boldsymbol{f} : I_T \times B_d \to \mathbb{R}^n$ is a continuous function, and denote the maximum of $||\boldsymbol{f}||$ by M. Suppose also that $\boldsymbol{f}$ satisfies a Lipschitz condition in $\boldsymbol{x} \in B_d$, with Lipschitz constant independent of t, i.e., there is a $k > 0$ such that*

$$\text{for all } t \in I_T \quad \boldsymbol{x}, \boldsymbol{y} \in B_d \quad ||\boldsymbol{f}(t, \boldsymbol{x}) - \boldsymbol{f}(t, \boldsymbol{y})|| \leq k||\boldsymbol{x} - \boldsymbol{y}||.$$

Then the initial value problem (5.21) *has a unique solution in the interval I_δ, with*

$$\delta = min\left\{T, \frac{d}{M}\right\}. \tag{5.24}$$

Sketch of Proof The chief difficulty in this proof compared to the one of Theorem 5.2 lies in ensuring that the map Γ in (5.22) takes $C(I_\delta, B_d)$ to itself, for suitably chosen $\delta > 0$. The problem is that, for $\boldsymbol{x} \in C(I_\delta, B_d)$, we may violate the condition

$$||\Gamma(\boldsymbol{x})(t) - \boldsymbol{x}_0|| \leq d \tag{5.25}$$

for some t. To see that one can always find a $\delta > 0$ so that this does not happen for $t \in I_\delta$, we note

$$||\Gamma(\boldsymbol{x})(t) - \boldsymbol{x}_0|| = ||\int_{t_0}^{t} \boldsymbol{f}(s, \boldsymbol{x}(s))\, ds|| \leq \int_{t_0}^{t} ||\boldsymbol{f}(s, \boldsymbol{x}(s)||\, ds \leq \delta M.$$

Hence the condition (5.25) is satsified provided we choose δ as in (5.24). □

Exercise 5.10 Complete the proof of Theorem 5.3.

In many applications it is useful to know the maximal interval on which the solution of an initial value problem exists and is unique. Perhaps not surprisingly, one can show that solutions can always be continued, in a unique fashion, until either the solution tends to infinity or reaches the boundary of the region $I_T \times B_d$ on which the function $\boldsymbol{f}$ satisfies the assumptions of Theorem 5.3. For a discussion of these matters, have a look at textbooks like Hale [1] or search for lectures notes on the Picard–Lindelöf Theorem on the internet.

References

[1] J. Hale, *Ordinary Differential Equations*, Dover Publications, 2009.

[2] P. E. Hydon, *Symmetry Methods for Differential Equations*, Cambridge University Press, Cambridge, 2000.

[3] D. W. Jordan and P. Smith, *Nonlinear Ordinary Differential Equations*, Third Edition, Oxford University Press, Oxford, 1999.

[4] S. J. Chapman, J. Lottes and L. N. Trefethen, Four Bugs on a Rectangle, *Proc. Roy. Soc. A*, **467** (2011), 881–896.

[5] B. J. Schroers, Bogomol'nyi Solitons in a Gauged O(3) Sigma Model, *Physics Letters B*, **356** (1995), 291–296; also available as an electronic preprint at http://xxx.soton.ac.uk/abs/hep-th/9506004.

[6] S. H. Strogatz, D. M. Abrams, A. McRobie, B. Eckhardt and E. Ott, Crowd Synchrony on the Millennium Bridge, *Nature*, **438** (2005), 43–44.

[7] M. Abrams, *Two coupled oscillator models: The Millennium Bridge and the chimera state*, PhD Dissertation, Cornell University, 2006.

[8] P. Dallard, A. J. Fitzpatrick, A. Flint, S. Le Bourva, A. Low, R. M. R. Smith and M. Willford, The Millennium Bridge London: Problems and Solutions, *The Structural Engineer*, **79:22** (2001), 17–33.

[9] M. Atiyah and N. Hitchin, *Geometry and Dynamics of Magnetic Monopoles*, Princeton University Press, Princeton, 1988.

Index